Bernd M. Rode,
Thomas S. Hofer,
and Michael D. Kugler
The Basics of Theoretical and Computational Chemistry

1807–2007 Knowledge for Generations

Each generation has its unique needs and aspirations. When Charles Wiley first opened his small printing shop in lower Manhattan in 1807, it was a generation of boundless potential searching for an identity. And we were there, helping to define a new American literary tradition. Over half a century later, in the midst of the Second Industrial Revolution, it was a generation focused on building the future. Once again, we were there, supplying the critical scientific, technical, and engineering knowledge that helped frame the world. Throughout the 20th Century, and into the new millennium, nations began to reach out beyond their own borders and a new international community was born. Wiley was there, expanding its operations around the world to enable a global exchange of ideas, opinions, and know-how.

For 200 years, Wiley has been an integral part of each generation's journey, enabling the flow of information and understanding necessary to meet their needs and fulfill their aspirations. Today, bold new technologies are changing the way we live and learn. Wiley will be there, providing you the must-have knowledge you need to imagine new worlds, new possibilities, and new opportunities.

Generations come and go, but you can always count on Wiley to provide you the knowledge you need, when and where you need it!

William J. Pesce
President and Chief Executive Officer

Peter Booth Wiley
Chairman of the Board

Bernd M. Rode, Thomas S. Hofer, and Michael D. Kugler

The Basics of Theoretical and Computational Chemistry

WILEY-VCH Verlag GmbH & Co. KGaA

The Authors

Prof. Dr. Bernd Michael Rode
University of Innsbruck
Institute of General, Inorganic,
and Theoretical Chemistry
Innrain 52a
6020 Innsbruck
Austria

Dr. Thomas Hofer
Institute of General, Inorganic,
and Theoretical Chemistry
University of Innsbruck
Innrain 52a
6020 Innsbruck
Austria

Michael Kugler
Institute of General, Inorganic,
and Theoretical Chemistry
University of Innsbruck
Innrain 52a
6020 Innsbruck
Austria

Library of Congress Card No.: applied for

British Library Cataloguing-in-Publication Data
A catalogue record for this book is available from the British Library.

Bibliographic information published by the Deutsche Nationalbibliothek
The Deutsche Nationalbibliothek lists this publication in the Deutsche Nationalbibliografie; detailed bibliographic data are available in the Internet at ⟨http://dnb.d-nb.de⟩.

Printed in the Federal Republic of Germany
Printed on acid-free paper

Typesetting Asco Typesetter, Hong Kong
Printing Strauss GmbH, Mörlenbach
Binding Litges & Dopf Buchbinderei GmbH, Heppenheim
Cover Design Adam-Design, Bernd Adam, Weinheim

ISBN: 978-3-527-31773-8

Contents

The Basics of Theoretical and Computational Chemistry. Edited by B. M. Rode, T. S. Hofer and M. D. Kugler

ISBN: 978-3-527-31773-8

Preface

Computational Chemistry has become an indispensable part of chemical research in nearly all fields and Theoretical Chemistry an increasingly prominent subject in modern chemistry curricula. This book provides an easy access to both topics for undergraduate students as well as for chemists who did not receive appropriate theoretical teaching when they studied, and need or wish to upgrade their knowledge in order to read contemporary publications or to make proper use of available commercial packages of computational chemistry in their work.

There are many good books available for these topics, but they usually take the classical approach to quantum mechanics and – besides being quite voluminous – require a considerable knowledge of higher mathematics. Other, less fastidious books sacrifice accuracy in the underlying physics for the sake of simplicity and perpetuate simple model pictures which are often not compatible with rigorous quantum mechanics. This book attempts to keep the level of theory high, while omitting all historical examples and avoiding the more complicated formalism of integro-differential equations by using the vector space concept. The contents of this book should form a good basis for further, specialised reading aims to prevent any unqualified 'black-box' use of quantum chemical calculation or simulation packages through stepwise explanation of all simplifications leading to the commonly used computational methods in chemistry. The reader will also understand which models employed in the interpretation of chemical processes are compatible with or in contradiction to the valid theory.

It is hoped, therefore, that students, chemistry teachers and research chemists in academia and industry will equally appreciate this concise presentation of the most relevant topics of theoretical and computational chemistry, from fundamental principles to practical methods.

The authors wish to express their gratitude to Margaret Ostermann for her careful reading of the text, and to Andreas Pribil for his valuable technical assistance in preparing the manuscript. Our thanks are also due to Dr. Martin Ottmar of Wiley-VCH Publishing Company who encouraged us to write this book.

Innsbruck, Austria
November 2006

Bernd M. Rode
Thomas S. Hofer
Michael D. Kugler

The Basics of Theoretical and Computational Chemistry. Edited by B. M. Rode, T. S. Hofer and M. D. Kugler

ISBN: 978-3-527-31773-8

1 Introduction

The book of nature you can only understand, if you have previously learnt its language and the letters. It is written in mathematical language and the letters are geometrical figures, and without these means it is impossible for human beings to understand even a word of it.

Galileo Galilei, 16th century

1.1 Theory and Models – Interpretation of Experimental Data

Every experimental result is – at the best – as good as the theoretical model used for its interpretation.... This sentence has been chosen as a leading remark for this book, as chemistry is still largely a field of science dominated by more or less simple models employed to 'explain' (rather: describe) the behaviour of molecules, their interactions and reactions, and to interpret all the data acquired by sophisticated measurements. The terminology introduced by these models is very rarely questioned for its validity and compatibility with state-of-the-art theoretical knowledge, and thus many of the interpretations of experimental data and subsequent conclusions could be inappropriate or even erroneous. On the other hand, a valid theoretical background for chemistry has been existent since 1926, when Erwin Schrödinger formulated his famous equation. The reason why theory has not yet penetrated chemistry as it did physics during the early 20th century is to be seen in the inability to solve Schrödinger's equation by analytical methods for other than one-electron systems. To date, only numerical solutions are possible, and one had to wait for the capacity of high-performance computers to deal with chemically relevant systems to test, prove, and successfully apply theoretical chemistry methods. This capability has been reached only recently – for a larger scientific community during the past two decades – and it can be expected, therefore, that the 21st century will be the one in which theory is given a much greater role in chemistry, reforming curricula and research practice similar to the development of physics some 100 years ago.

The Basics of Theoretical and Computational Chemistry. Edited by B. M. Rode, T. S. Hofer and M. D. Kugler

ISBN: 978-3-527-31773-8

Today, computational chemistry methods are almost ubiquitously used, and publications frequently contain 'theoretical' sections which are often nothing more than black-box applications of commercial program packages. Such an inappropriate use of theoretical methods is often caused by a lack of knowledge of the quantum theoretical foundations of the available programs and a simultaneous adherence to simplistic models employed to interpret the quantitative results obtained by the computations in qualitative terms of these models.

The purpose of this book is to familiarise not only chemistry students, but also those chemists who did not have the chance to obtain a good theoretical background during their studies, with the basics of theoretical and computational chemistry. To an extent, this should enable them to understand the underlying physical principles, to judge the validity of the commonly used models, and to obtain sufficient knowledge to use computational chemistry methods in a prudent and appropriate way. In order to achieve these goals, mathematical requirements have been reduced to a minimum, without sacrificing physical rigour in the theoretical framework. This could be realised by using the vector space theory of matter with its largely linear algebraic formalism instead of the commonly employed formalism of integro-differential equations following the historical development of quantum chemistry. Detailed descriptions of procedures that might be important for the specialist – but not for the generally interested reader – have been omitted for the sake of clarity and conciseness. In the final sections, the basic principles of perturbation theory and group theory are outlined with regard to their use in chemistry, and a brief overview of the most important methods in contemporary computational chemistry is provided. Thus, the book should serve as a good general introduction into theory of chemistry and create a good basis for further, more specialised reading, wherever this is needed or desired.

Before starting the first chapters of this book, it seems important to attempt a definition of the difference between 'models' and 'theory'. A 'model' can be considered a fiction with a certain power to indicate some aspects of 'reality' and to rationalise a (limited) number of observations, sometimes also allowing some predictions. A 'theory' should allow a general and precise explanation/description of all (or at least as many as possible) phenomena and a reliable prediction of the results of any future observation. In this case, the fewer postulates needed in its formalism and the wider its field of validity, the better the theory. Chemistry is in the somewhat fortunate situation of being mainly concerned with very few of the manifold elementary particles, namely with electrons, nuclei (which can mostly be treated as charged masses of virtually no dimension), and photons. During the course of this book it will be seen that a theoretical concept for this case can be built on three well-proven postulates and two empirical observations, and that all further aspects of the theory result as logical consequences, thus making the development of the theory a straightforward procedure. Figure 1.1 provides a schematic overview of the general concept followed in this book.

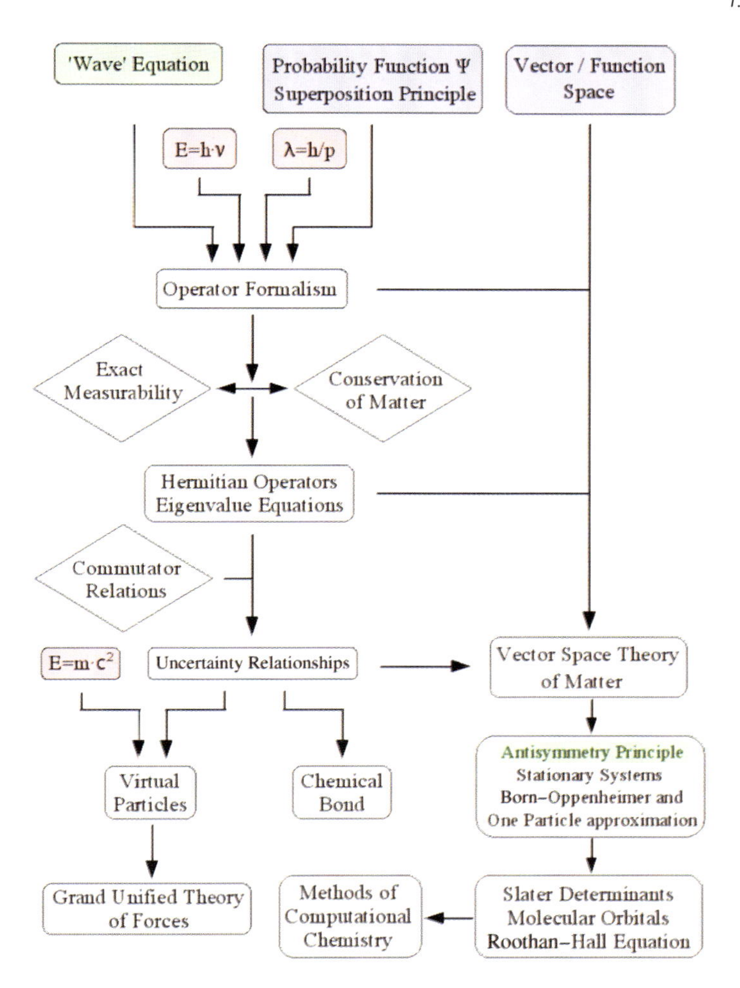

Fig. 1.1 'Float Chart' of the concept followed by this book. Postulates are indicated by red, empirical observations by green, and mathematical concepts by blue.

1.2
The Notation

To facilitate reading of this book, the consistent notation used in it will be summarised here, defining symbols and associated meanings. The following list should also serve as a convenient tool to identify any specific notation used in the text.

$|x\rangle$ denotes a vector in the 'ket' space, $\langle x|$ denotes the complex conjugated and transposed of $|x\rangle$, and thus a vector of the 'bra' space.

The associated sets of vectors are called 'bra' and 'ket', because $\langle\ \rangle$ corresponds to a 'bracket'. This notation is also known as *Dirac* notation. In detail, the vectors are defined as follows:

$$|x\rangle = (x_1, x_2, \ldots) \qquad \langle x| = \begin{pmatrix} x_1^* \\ x_2^* \\ \ldots \end{pmatrix}$$

φ denotes a function, and is usually also written as a vector $|\varphi\rangle$, due to the essential equivalence of a function with a vector: if we write the values of a function at subsequent values of its variable, we obtain an n-tuple of numbers, i.e., a vector ('digitised function').

$\mathbb{A}$ or $\mathbb{S}$ ('fat symbols') are used for matrices, and this is equally applied to Greek letters: $\boldsymbol{\Phi}$ and $\boldsymbol{\Psi}$ denotes matrices.

The determinant of a matrix is given by $|\mathbb{A}| = \det(\mathbb{A})$. The transposed matrix $\mathbb{A}$ is written as $\mathbb{A}^T$, the adjugate matrix as $\mathbb{A}^+$ and the inverse as $\mathbb{A}^{-1}$.

$$\mathbb{1} = \begin{pmatrix} 1 & \cdots & \emptyset \\ \emptyset & \cdots & \emptyset \\ \cdots & \cdots & \cdots \\ \emptyset & \cdots & 1 \end{pmatrix}$$

is the unit matrix and $\mathbb{0}$ the zero-matrix (containing only $\emptyset$ values).

Vectors and functions are often constructed from sets of more or less simple 'basis' vectors/functions. $\{e_i\}^n$ and $\{\varphi_i\}^n$ indicate such basis sets.

Operators will be marked with a hat; for example, $\hat{H}$ stands for the Hamilton operator. Special operators are the direct-sum $\oplus$ and the direct-product $\otimes$ operators and the $\hat{1}$ (identity) and $\hat{0}$ (annihilation) operators.

Two other frequently used special operators are the Nabla and the Laplace operator:

$$\text{Nabla operator:} \quad \nabla = \frac{\partial}{\partial x}|i\rangle + \frac{\partial}{\partial y}|j\rangle + \frac{\partial}{\partial z}|k\rangle$$

$$\text{Laplace operator:} \quad \Delta = \frac{\partial^2}{\partial x^2} + \frac{\partial^2}{\partial y^2} + \frac{\partial^2}{\partial z^2}$$

The Nabla operator is a vector operator, while the Laplace operator is scalar, resulting from the scalar product of the Nabla Operator with itself: $\Delta = \langle\nabla|\nabla\rangle$.

The Kronecker–Delta is another important notation, meaning:

$$\delta_{ij} = \begin{cases} 1 & \forall i = j \\ \emptyset & \text{otherwise} \end{cases}$$

The following symbols are also used in this book with the meanings:

$\exists$ ... there is

$\forall$ ... for all

$\in$ ... is an element of

iff ... if and only if

$d\tau = dx\,dy\,dz$

1.3 Vector Space V^n and Function Space F^n

V^n denotes a vector space of the dimension n. Thus, V^2 is a two-dimensional vector space (i.e., a plane) and V^3 is a three-dimensional vector space, corresponding to our familiar x/y/z coordinate system. To describe a vector space, one needs 'basis vectors': n linearly independent vectors span an n-dimensional vector space. Linear independence means that none of these basis vectors can be constructed by a linear combination of the other n − 1 basis vectors. If all basis vectors are perpendicular to each other, they form an orthogonal basis set.

Basis sets are denoted as $\{e_i\}^n$ for V^n. In Cartesian coordinates the basis for the V^3 consists of three basis vectors in x-, y-, and z-direction:

$$\{e_i\}^3 = (|i\rangle, |j\rangle, |k\rangle)$$

The function space F^n is the analogue of the vector space, using n linearly independent functions φ_i as basis $\{\varphi_i\}^n$.

Basically, a function is a transformation from the variable x to its function value $f(x)$. In fact it is a continuous transformation, but if one digitizes the function in steps Δx, a finite vector is obtained instead of the function. So, x is reproduced as f_1, $x + \Delta x$ as f_2 and so on, resulting in an n-tuple of numbers $(f_1, f_2, \ldots)$. If Δx decreases toward zero, the dimension of the vectors grows, finally reaching infinity, i.e., a vector in V^∞.

The equivalence between vector and function space will be used throughout this book, as it allows a much more convenient and easier-to-manipulate formulation of equations, which can easily be transformed into computer language.

The Scalar Product

Given, the vectors $|a\rangle$ and $|b\rangle$, the scalar product $\langle a|b\rangle$ is obtained by $\sum_i a_i^* \cdot b_i$. The analogue for functions is a sum with infinitesimal step width, i.e., an integral. Thus, $\langle\varphi_1|\varphi_2\rangle$ corresponds to $\int \varphi_1^* \cdot \varphi_2 \, d\tau$. We will make continuous use of this simplified notation of integrals in function space as scalar products in vector space.

1.4
Linear Transformation – Change of Basis

Often it is an advantage to change bases. To change from a basis $\{e_i\}^n = (e_1 \dots e_n)$ to another basis of the same dimension n, $\{x_i\}^n = (x_1 \dots x_n)$, one performs a linear tranformation. As an example we will consider a vector $|u\rangle$, represented in both bases.

Let $|u\rangle$ be an element of V^2, given in $\{e_i\}^2$ as $|u\rangle^e = 3|e_1\rangle + 2|e_2\rangle$. We wish to obtain its representation in $\{x_i\}^2$, namely $|u\rangle^x$. For this purpose, we must know the relationship between the two bases. As every vector of V^2 can be constructed from a given basis set, the two vectors forming the basis $\{x_i\}^2$ can also be constructed from the basis vectors of $\{e_i\}^2$. In our example, let the connection between the different basis vectors be given by:

$$|x_1\rangle = 2|e_1\rangle - |e_2\rangle \qquad \text{and} \qquad |x_2\rangle = |e_1\rangle + 3|e_2\rangle.$$

In matrix notation this reads $\mathbb{X} = \mathbb{E} \cdot \begin{pmatrix} 2 & 1 \\ -1 & 3 \end{pmatrix} = \mathbb{E} \cdot \mathbb{T}$, or $\mathbb{X}\mathbb{T}^{-1} = \mathbb{E}$, where $\mathbb{T}$ is the transformation matrix from $\mathbb{E}$ to $\mathbb{X}$.

We can now write

$$\mathbb{U} = \mathbb{E}\begin{pmatrix} 3 \\ 2 \end{pmatrix} = \mathbb{X}\mathbb{T}^{-1}\begin{pmatrix} 3 \\ 2 \end{pmatrix}$$

with $\mathbb{T}^{-1} = \begin{pmatrix} \frac{3}{7} & -\frac{1}{7} \\ \frac{1}{7} & \frac{2}{7} \end{pmatrix}$, we obtain $\mathbb{U} = \mathbb{X} \cdot \begin{pmatrix} 1 \\ 1 \end{pmatrix}$ or $|u\rangle^x = |x_1\rangle + |x_2\rangle$. This small example demonstrates that we only have to know the transformation matrix between two bases to calculate the coefficients of a vector given in one basis in the other one.

1.5
Normalisation and Orthogonalisation of Vectors

A vector is called normalised, if the scalar product with itself equals 1, and when the scalar product of two different vectors yields zero, they are called orthogonal to each other. If a set of vectors fulfils the condition $\langle e_i|e_j\rangle = \delta_{ij}$, they form an orthonormalised or unitary vector system, respectively. To normalise a vector $|u\rangle$, one calculates $\frac{|u\rangle}{\sqrt{\langle u|u\rangle}} = \frac{|u\rangle}{|u|} = |e_u\rangle$, where $|e_u\rangle$ results as a normalised or 'unit' vector in the direction of $|u\rangle$. Orthogonalisation of basis vectors can be achieved by a stepwise procedure ('Schmidt Orthogonalisation'), but there is a very convenient way to orthonormalise a complete basis set at once applying the *Löwdin Orthonormalisation.*

Let us note the basis $\{x_i\}$ as matrix $\mathbb{X}$. $\langle \mathbb{X}|\mathbb{X}\rangle = \boldsymbol{\Delta}$ and is called the metrix of $\mathbb{X}$. The metrix is defined by:

$$\boldsymbol{\Delta} = \begin{pmatrix} \langle x_1|x_1\rangle & \langle x_1|x_2\rangle & \cdots & \langle x_1|x_n\rangle \\ \langle x_2|x_1\rangle & \cdots & \cdots & \cdots \\ \cdots & \cdots & \cdots & \cdots \\ \langle x_n|x_1\rangle & \cdots & \cdots & \langle x_n|x_n\rangle \end{pmatrix}$$

We now form a new basis matrix $\mathbb{X}'$ by multiplying the original basis by $\boldsymbol{\Delta}^{-\frac{1}{2}}$:

$$\mathbb{X} \cdot \boldsymbol{\Delta}^{-\frac{1}{2}} = \mathbb{X}'$$

It is easily shown that the new basis set $\{x_i'\}$ consists of orthonormalised vectors:

$$\begin{aligned} \langle \mathbb{X}'|\mathbb{X}'\rangle &= \langle \boldsymbol{\Delta}^{-\frac{1}{2}} \cdot \mathbb{X}|\mathbb{X} \cdot \boldsymbol{\Delta}^{-\frac{1}{2}}\rangle = \boldsymbol{\Delta}^{-\frac{1}{2}}\langle \mathbb{X}|\mathbb{X}\rangle\boldsymbol{\Delta}^{-\frac{1}{2}} = \boldsymbol{\Delta}^{-\frac{1}{2}} \cdot \boldsymbol{\Delta} \cdot \boldsymbol{\Delta}^{-\frac{1}{2}} \\ &= \boldsymbol{\Delta}^{+\frac{1}{2}} \cdot \boldsymbol{\Delta}^{-\frac{1}{2}} = \mathbb{1} \end{aligned}$$

It will prove very convenient to use orthonormalised basis sets in many cases. With the help of the Löwdin procedure, we can always convert any primary choice of basis set into an orthonormalised one.

1.6 Matrix Representation of the Scalar Product

The scalar product of two vectors is given by

$$\langle u|v\rangle = \sum_i u_i^* \cdot v_i = \mathbb{U}^+ \cdot \mathbb{V}$$

The latter expression corresponds to the matrix notation, and it is evident that the matrices $\mathbb{U}$ and $\mathbb{V}$ have to correspond to the same basis $\{e_i\}$. Explicitly, the vectors are given as:

$$\begin{aligned} |u\rangle &= c_{u_1} \cdot |e_1\rangle + c_{u_2} \cdot |e_2\rangle + \cdots + c_{u_n} \cdot |e_n\rangle = \mathbb{E} \cdot \mathbb{C}_u \\ |v\rangle &= c_{v_1} \cdot |e_1\rangle + c_{v_2} \cdot |e_2\rangle + \cdots + c_{v_n} \cdot |e_n\rangle = \mathbb{E} \cdot \mathbb{C}_v \end{aligned}$$

where $\mathbb{E}$ represents the basis, and the coefficients $(c_{u_1}, c_{u_2}, \ldots, c_{u_n})$ of $|u\rangle$ form the column vector $\mathbb{C}_u$, the coefficients of $|v\rangle$ the column vector $\mathbb{C}_v$. Thus, the scalar product of $|u\rangle$ and $|v\rangle$ is given by:

$$\langle u|v\rangle = \mathbb{U}^{+} \cdot \mathbb{V} = \mathbb{C}_u^{+}\langle\mathbb{E}|\mathbb{E}\rangle\mathbb{C}_v = \mathbb{C}_u^{+}\mathbf{\Delta}\mathbb{C}_v$$

or simply by $\mathbb{C}_u^{+}\mathbb{C}_v$, if the basis is orthonormalised.

The same formalism is valid in function space. Instead of $\{e_i\}$ we use $\{\varphi_i\}$

$$\psi_u = \mathbf{\Phi} \cdot \mathbb{C}_u$$
$$\psi_v = \mathbf{\Phi} \cdot \mathbb{C}_v$$

The scalar product $\langle\psi_u|\psi_v\rangle$ – corresponding to integration – is thus given as $\mathbb{C}_u^{+}\langle\mathbf{\Phi}|\mathbf{\Phi}\rangle\mathbb{C}_v$ with $\langle\mathbf{\Phi}|\mathbf{\Phi}\rangle = \mathbb{S}$, the metrix, which is called the overlap integral matrix in function space.

$$\mathbb{S} = \begin{pmatrix} \langle\varphi_1|\varphi_1\rangle & \cdots & \langle\varphi_1|\varphi_n\rangle \\ \cdots & & \\ \langle\varphi_n|\varphi_1\rangle & \cdots & \langle\varphi_n|\varphi_n\rangle \end{pmatrix}$$

$$\mathbb{S} = \begin{pmatrix} \int\varphi_1^*\varphi_1\,d\tau & \cdots & \int\varphi_1^*\varphi_n\,d\tau \\ \cdots & & \\ \int\varphi_n^*\varphi_1\,d\tau & \cdots & \int\varphi_n^*\varphi_n\,d\tau \end{pmatrix}$$

if $\langle\varphi_i|\varphi_j\rangle = \delta_{ij}$, then $\mathbb{S} = \mathbb{1}$.

The Löwdin orthonormalisation can be analogously performed in function space: starting from an arbitrary basis $\{\varphi_i\}$ one obtains $\langle\mathbf{\Phi}|\mathbf{\Phi}\rangle = \mathbb{S}$, and transforming the basis by $\mathbf{\Phi}' = \mathbf{\Phi} \cdot \mathbb{S}^{-\frac{1}{2}}$, one has $\langle\mathbf{\Phi}'|\mathbf{\Phi}'\rangle = \mathbb{1}$, which means that $\mathbf{\Phi}'$ is the desired orthonormalised basis.

1.7
Dual Vector Space and Hilbert Space

While talking of scalar products, we have silently assumed that we can find a transposed, complex conjugated form for every vector as well, thus implying the condition:

$$\exists\langle a|\forall|a\rangle \in V^n$$

A vector space fulfilling this condition is called a *Dual Vector Space*. This condition is essential for a vector space describing a physical system, as the evaluation of physical quantities implies the formation of scalar products. In addition to that, two further conditions will be imposed on the vector space we are going to use for the description of matter, namely the two *Cauchy convergence criteria*:

$$\sum_{i=1}^{\infty}\langle\Psi_i|\Psi_i\rangle < \infty$$

and

$$\lim_{M,N\to\infty} \sum_n |\Psi_n^M - \Psi_n^N|^2 = \emptyset$$

employing different representations of Ψ.

These criteria are also termed 'quadratic convergence' of the vector space/function space, and a dual vector space fulfilling them is called a *Hilbert Space*. In the next section we will see that quadratic convergence is required in order to guarantee the normalisability of a system to a finite number of particles.

1.8 Probability Concept and the Ψ Function

The probability concept implies that we can describe any physical system by a function/vector containing all the information about this system, i.e. its properties as a function of coordinates, and that we can evaluate the probability for any state of the system from this function by the product $\Psi^*\Psi$. The overall probability is then given by the scalar product

$$\int_{-\infty}^{\infty} \Psi^* \cdot \Psi \, d\tau = \langle \Psi | \Psi \rangle$$

It is now understood, why quadratic convergence of the vector space to which $|\Psi\rangle$ belongs is required, as otherwise the overall probability could not be restrained to a finite value.

The probability function is usually formulated as a function of space coordinates and the time coordinate as $\Psi(r_i, t)$, but in some cases the use of momentum coordinates ('momentum space') and time as $\Psi(|p_i\rangle, t)$ is advantageous.

The concept describing a physical system by a probability function/vector depending on a few variables implies that other physical variables must be evaluated from this function/vector. The mathematical instruments achieving this are called *operators*, and the general definition and properties of such operators will be detailed in the next section.

1.9 Operators

In this textbook, an operator is marked with a 'hat', (e.g., $\hat{H}$). The operator's action in vector space can be illustrated, if we consider the vector $|x\rangle$ and an operator $\hat{A}$: $\hat{A}|x\rangle = |x'\rangle$, i.e., the operator transforms the original vector, in general by changing its amount and its direction, into a new vector. One also says that the original vector is mapped onto an image vector. Some examples of operators are

$\widehat{C}_2$, representing a rotation by 180 degrees and $\widehat{C}_\varphi$, corresponding to a rotation by the angle φ. Another example is the inversion operator $\hat{i}$, by which every component of the vector is mapped onto its inverse, such as $x \to -x$. In V^3 the matrix $\mathbb{I}$ representing this operator is:

$$\mathbb{I} = \begin{pmatrix} -1 & 0 & 0 \\ 0 & -1 & 0 \\ 0 & 0 & -1 \end{pmatrix}$$

As the inversion operator and this matrix are associated by $\hat{i}\mathbb{X} = \mathbb{X}\mathbb{I}$, every operator can be represented in matrix form, once a vector space has been defined by a concrete basis.

It will be useful to summarise a few basic definitions and rules for operators. There are two types of operator: (i) *regular* operators, for which a one-to-one correspondence between original vectors and image vectors exists; and (ii) *singular* operators, for which the correspondence is only unidirectional from original vectors to image vectors. The previously given inversion operator is an example of a regular operator, the square operator $(\ )^2$ is a singular operator, as it is not possible to distinguish from a value 4 whether the original value in the domain of the operator was 2 or -2 (the set of all original vectors is called its 'domain').

An *inverse* operator multiplied with its source operator results in the *unity* operator, according to $\hat{A}^{-1}\hat{A} = \hat{1}$.

In general, operators are not commutative, i.e., $\hat{A}\hat{B} \neq \hat{B}\hat{A}$.

Linear operators are defined by the following conditions:

- $\hat{A}(|u\rangle + |v\rangle) = \hat{A}|u\rangle + \hat{A}|v\rangle$
- $\hat{A}(k|u\rangle) = k\hat{A}|u\rangle$, k being a scalar
- $(\widehat{A_1} + \widehat{A_2})|u\rangle = \widehat{A_1}|u\rangle + \widehat{A_2}|u\rangle$

Most of the operators we are using in vector space theory of matter are linear operators.

1.10 Representation of Operators in a Basis

In a given basis $\{e_i\}^n$, an operator will act on each of the basis vectors by transforming it into another vector, which will also be an element of the same vector space $\hat{A}|e_1\rangle = |e'\rangle \in V^n$, and can be constructed, therefore, as a linear combination of all original basis vectors:

$$\hat{A}|e_1\rangle = a_{11}|e_1\rangle + a_{12}|e_2\rangle + \cdots + a_{1n}|e_n\rangle = \sum_i a_{1i}|e_i\rangle$$

$$\hat{A}|e_2\rangle = \sum_i a_{2i}|e_i\rangle, \quad \text{until } \hat{A}|e_n\rangle = \sum_i a_{ni}|e_i\rangle$$

The first index of $a_{1i}, a_{2i}, \ldots a_{ni}$ specifies which basis vector is transformed. All of these equations can be shortly summarised in matrix form as

$$\hat{A}\mathbb{E} = \mathbb{E}\mathbb{A}^e \tag{1}$$

where $\mathbb{A}^e$ contains all coefficients a_{ij} and is thus a complete representation of the operator in the basis $\{e_i\}^n$. If Eq. (1) is multiplied with the basis in the bra space, one obtains: $\langle\mathbb{E}|\hat{A}\mathbb{E}\rangle = \langle\mathbb{E}|\mathbb{E}\rangle\mathbb{A}^e$ and from this

$$\mathbb{A}^e = \frac{\langle\mathbb{E}|\hat{A}\mathbb{E}\rangle}{\langle\mathbb{E}|\mathbb{E}\rangle}$$

which is the general expression for the representation of the operator in the given basis. This expression is also termed the *expectation value* of the operator $\hat{A}$. For orthonormal bases, this reduces to $\mathbb{A}^e = \langle\mathbb{E}|\hat{A}\mathbb{E}\rangle$.

The matrix $\mathbb{A}^e$ contains the following elements:

$$\mathbb{A}^e = \begin{pmatrix} \langle e_1|\hat{A}e_1\rangle & \langle e_1|\hat{A}e_2\rangle & \cdots & \langle e_1|\hat{A}e_n\rangle \\ \langle e_2|\hat{A}e_1\rangle & \cdots & \cdots & \cdots \\ \cdots & \cdots & \cdots & \cdots \\ \langle e_n|\hat{A}e_1\rangle & \cdots & \cdots & \langle e_n|\hat{A}e_n\rangle \end{pmatrix}$$

In function space the analogous representation of an operator in a basis of functions is given as

$$\mathbb{A}^\varphi = \frac{\langle\boldsymbol{\Phi}|\hat{A}\boldsymbol{\Phi}\rangle}{\langle\boldsymbol{\Phi}|\boldsymbol{\Phi}\rangle} \quad \text{or} \quad \langle\boldsymbol{\Phi}|\hat{A}\boldsymbol{\Phi}\rangle$$

for an orthonormalised basis set. In detail, this matrix consists of the elements:

$$\mathbb{A}^\varphi = \begin{pmatrix} \int \varphi_1^*\hat{A}\varphi_1\,d\tau & \cdots & \int \varphi_1^*\hat{A}\varphi_n\,d\tau \\ \cdots & \cdots & \cdots \\ \int \varphi_n^*\hat{A}\varphi_1\,d\tau & \cdots & \int \varphi_n^*\hat{A}\varphi_n\,d\tau \end{pmatrix}$$

In order to recall the correspondence between scalar products and integrals, the elements of this matrix have been written as integrals.

1.11
Change of Basis in Representations of Operators

If we have two different bases $\{e_i\}$ and $\{f_i\}$, whose connection is given by $\mathbb{F} = \mathbb{E}\mathbb{T}$ (linear transformation), we can connotate the equation $\hat{A}|u\rangle = |v\rangle$ as $\mathbb{V}^e = \mathbb{A}^e\mathbb{U}^e$ or $\mathbb{V}^f = \mathbb{A}^f\mathbb{U}^f$, depending on the chosen basis.

Based on $\mathbb{U}^e = \mathbb{T}\mathbb{U}^f$ and $\mathbb{V}^e = \mathbb{T}\mathbb{V}^f$ we can formulate $\mathbb{T}\mathbb{V}^f = \mathbb{A}^e\mathbb{T}\mathbb{U}^f$, which upon multiplication with $\mathbb{T}^{-1}$ from the left leads to

$$\mathbb{V}^f = \mathbb{T}^{-1}\mathbb{A}^e\mathbb{T}\mathbb{U}^f = \mathbb{A}^f\mathbb{V}^f,$$

from which follows

$$\mathbb{A}^f = \mathbb{T}^{-1}\mathbb{A}^e\mathbb{T}$$

Such a transformation is called *similarity transformation*, and if $\mathbb{T}^{-1} = \mathbb{T}^+$, *unitary transformation*.

Changing the basis (e.g., a coordinate system) must not change the physics of the system. Besides the expectation value, two further properties of a matrix are invariant in similarity transformations, namely the determinant and the trace of the matrix:

$$det(\mathbb{A}) = det(\mathbb{T}^{-1}\mathbb{A}\mathbb{T})$$

$$tr(\mathbb{A}) = \sum_i a_{ii} = tr(\mathbb{T}^{-1}\mathbb{A}\mathbb{T})$$

Consequently, all physical quantities associated with the operator will have to be associated with its expectation value, the determinant, and/or trace of its representation matrix.

Test Questions Related to this Chapter

1. Why can we treat functions as vectors?
2. What are the differences between a model and a theory?
3. Why does vector space designed to describe matter have to be a Hilbert space?
4. Can we use the probability concept in classical physics?
5. What possibilities exceeding classical physics offer the probability concept in the description of particle behaviour?
6. What consequences does the probability concept have for the evaluation of physical variables?
7. What is a unitary transformation, and which characteristics of a matrix are invariant to it?

2
Basic Concepts of Vector Space Theory of Matter

The origin of all are the empty space and the atoms, everything else is just a belief.

Demokritos, 5th century BC

When scientists first began to investigate the properties of light, they employed two concepts. The first concept, proposed by *Newton*, described light as small particles (corpuscula), while the other concept, developed by *Huygens*, considered light as a wave, similar to the waves that propagate in water when a stone is thrown into it. This *'dual nature'* of light remained a major topic in physics until the 20th century and led to a search for an 'ether', in which the light waves could propagate. However, when the existence of such an ether was excluded by sophisticated experiments, another interpretation began to evolve, namely the idea that the 'wave' equation might simply be an expression of a specific behaviour of the small particles – the photons – of which light is composed. 'Wave-like' behaviour such as diffraction patterns was also observed for other particle beams (e.g., electron beams), and it was understood that every particle had an associated 'wave length', describing its behaviour in experiments directed at this 'wave nature'. For some time, the assumption of 'travelling' wave packages (thus avoiding the 'ether') became popular, until it was shown that such packages would disperse in the course of time and thus the behaviour as particles would be lost. It was to a large extent the merit of *Born* to identify the probabilistic nature of elementary particle behaviour, thus reducing the *'particle-wave dualism'* to a statistical interpretation of behaviour and interaction of particles. We will follow this concept and see the 'wave function' and the equations derived from the treatment of matter as waves as a mathematical expression of this probabilistic behaviour, thus arriving easily at the most important expressions and formulations of quantum theory.

2.1
The Wave Equation as Probability Function

We will now consider the well-known solutions for the wave equation as a proper quantitative description of the probabilistic behaviour of small particles such as

The Basics of Theoretical and Computational Chemistry. Edited by B. M. Rode, T. S. Hofer and M. D. Kugler

ISBN: 978-3-527-31773-8

photons or electrons, and take these solutions as the basis for our further development of the concept. The 'wave function' is, therefore, assumed to represent a probability function (or vector), which will allow us to evaluate physical observables with regard to the probability of values they can assume.

For the sake of simplicity it is convenient to restrict the formalism to one space dimension, in this case the x direction, and time. The solution for the (empirically derived) wave equation for light is given as

$$\psi(x,t) = C \cdot e^{i\alpha}$$

where the exponent α represents the so-called 'phase factor'

$$\alpha = 2\pi\left(\frac{x}{\lambda} - \nu \cdot t\right)$$

and λ is the 'wave length' associated with the particle, and ν the frequency, given by $\frac{c}{\lambda}$, c being the velocity of light.

2.2
The Postulates of Quantum Mechanics

At this point it becomes necessary to introduce the fundamental postulates of quantum mechanics. While trying to describe the radiation of a black body by a suitable mathematical formalism during the late 19th century, it was found by *Planck* that a correct formalism could only be found with the assumption that energy cannot be transmitted in a continuous way, but only in discrete amounts, for which the term 'quanta' was introduced. This postulate was in clear contradiction to classical physical thinking, and is the main pillar of quantum mechanics. The famous relationship associating the energy with the frequency and Planck's constant is

$$E = h\nu.$$

Subsequently, it was found that almost all physical variables obey similar laws – that is, they vary in discrete quantities and that this 'quantisation' is a fundamental property of matter. Recently, it has been found that even time seems to be quantised. The quanta carry names specific to the variable or property of matter with which they are associated: photons represent light, and the (more hypothetical) gravitons and chronons the gravitation force and time, respectively.

The second important postulate of quantum mechanics was derived from the famous energy–mass relationship:

$$E = mc^2$$

which was first formulated by *Hasenöhrl*, but is now generally attributed to *Einstein*. This relation has been verified in many ways, be it the formation of two elementary particles (electron and positron) from a γ–ray of sufficient energy or the mass defect going from heavy elements (e.g., U) or light elements (e.g., H) to those of medium masses, as observed in nuclear fission and fusion, respectively. Mass can be seen, therefore, as a special form of energy, and energy as particles of a given mass and momentum.

Comparing the two expressions for energy, one now obtains:

$$E_{Planck} = h \cdot \nu = m \cdot c^2 = E_{Einstein}$$

This can be resolved further to

$$mc^2 = h\nu = h\frac{c}{\lambda} \quad \rightarrow \quad mc = \frac{h}{\lambda} \quad \rightarrow \quad \lambda = \frac{h}{|p\rangle}$$

where mc is the momentum $|p\rangle$ of the photon. From this point we easily arrive at the second postulate of quantum mechanics, by extending this relationship to all particles, as proposed by *De Broglie*. Then, any particle with a given mass and speed has via its momentum $|p\rangle$ an associated wavelength λ, which – in the light of the probability concept – is related to the probabilistic behaviour of this particle. Basically, de Broglie's postulate means a generalisation to particles moving with a velocity v less than the speed of light c, resulting thus in

$$\lambda = \frac{h}{mv}$$

which is the second postulate we will have to employ in our formalism.

2.3 The Schrödinger Equation

If one introduces the two postulates of quantum mechanics into the phase factor

$$\alpha = 2\pi\left(\frac{x}{\lambda} - \nu \cdot t\right)$$

and makes use of this expression in the one-dimensional 'wave' equation

$$\psi(x,t) = C \cdot e^{i\alpha}$$

one obtains the following expression for the probability behaviour of particles:

$$\psi(x,t) = C \cdot e^{i \cdot 2\pi\left(\frac{xp_x}{h} - \frac{Et}{h}\right)} = C \cdot e^{\frac{i}{\hbar}(xp_x - Et)} \tag{2}$$

Differentiating this expression by time one obtains:

$$\frac{\partial \psi}{\partial t} = -\frac{i}{\hbar} \cdot E \cdot \psi$$

which can be rearranged as

$$i\hbar \frac{\partial \psi}{\partial t} = E\psi \tag{3}$$

This is the *time-dependent Schrödinger equation*, one of the fundamental equations of quantum mechanics.

Differentiation of Eq. (2) by x leads to

$$\frac{\partial \psi}{\partial x} = \frac{i}{\hbar} |p_x\rangle \psi$$

which is usually written as

$$-i\hbar \frac{\partial}{\partial x} \psi = |p_x\rangle \psi \tag{4}$$

On examining Eqs. (3) and (4), we can see that we have obtained two important operators delivering corresponding variables from the probability function, namely

$$i\hbar \frac{\partial}{\partial t} = \hat{E}$$

the operator for the energy, and

$$-i\hbar \frac{\partial}{\partial x} |i\rangle = \widehat{p_x}$$

the operator of the momentum in the x direction (the unit vector in the x direction $|i\rangle$ has been included in this operator to emphasise that this is a vectorial operator). The analogous operators to $\widehat{p_x}$ in y and z directions are

$$-i\hbar \frac{\partial}{\partial y} |j\rangle = \widehat{p_y}$$

$$-i\hbar \frac{\partial}{\partial z} |k\rangle = \widehat{p_z}$$

and for the operator of the total momentum $\hat{p}$ we thus obtain

$$-i\hbar\left(\frac{\partial}{\partial x}|i\rangle + \frac{\partial}{\partial y}|j\rangle + \frac{\partial}{\partial z}|k\rangle\right)$$

or, with the Nabla operator,

$$\hat{p} = -i\hbar\nabla$$

With these operators we can now build other operators for our vector space and the probability function, according to the *correspondence principle*, which states that there is a corresponding operator in quantum mechanics for every variable in classical mechanics. However, one should note that the reverse is not true: there are operators (and thus properties of matter) in quantum theory, which do not have any corresponding variable in classical physics. Before proceeding further we should also know that the variables of the probability functions are at the same time their own operators, in our function for example $\hat{x} = x$.

With the operators already derived we can now build the operator for the amount of momentum

$$\hat{p}^2 = -\hbar^2 \cdot \langle\nabla|\nabla\rangle = -\hbar^2\Delta$$

where Δ is the Laplace operator. For the kinetic energy we have the classical expression

$$T = E_{kin} = \frac{mv^2}{2} = \frac{p^2}{2m}$$

which leads to the operator for the kinetic energy

$$\hat{T} = -\frac{\hbar^2}{2m}\Delta$$

The total energy of a classical system is defined by the Hamilton function as the sum of kinetic and potential energy $H = T + V$. The corresponding *Hamilton operator* has the form

$$\hat{H} = -\frac{\hbar^2}{2m}\Delta + \hat{V}$$

and thus we can formulate the *time-independent Schrödinger equation*:

$$\hat{H}|\psi\rangle = E|\psi\rangle \tag{5}$$

Then, combining this with the time-dependent equation [Eq. (3)] we finally arrive at the *time-dependent Schrödinger equation in operator form*:

Fig. 2.1 The grave of Erwin Schrödinger in the churchyard of Alpbach, 60 km from Innsbruck, Austria.

$$i\hbar\frac{\partial}{\partial t}|\psi\rangle = \hat{H}|\psi\rangle \tag{6}$$

which often is also written as

$$i\hbar|\dot{\psi}\rangle = \hat{H}|\psi\rangle \tag{7}$$

We also find the equation in this form on Erwin Schrödinger's grave in Alpbach, near Innsbruck in Austria (Fig. 2.1).

In this chapter we have encountered a typical form of equations, namely operator · function = value · function, meaning that an operator acts on the probability function (or vector), producing thereby the value of the corresponding variable multiplied by the function/vector

$$\hat{O}|\psi\rangle = o|\psi\rangle$$

Such equations are called eigenvalue equations, the corresponding functions/vectors the eigenfunctions/eigenvectors of the operator $\hat{O}$, and the values o the eigenvalues of the variable associated with $\hat{O}$.

2.4 Hermicity

We have already stated that the probability w is calculated from the product of the probability function with its complex conjugate $w = \psi^*\psi$, and the total probability of a system is, therefore, given by integration over the whole configuration space as

$$W = \int_{-\infty}^{\infty} \psi^* \psi \, d\tau$$

Changes of the probability in time are found through differentiation of these expressions as

$$\frac{dw}{dt} = \frac{d\psi^* \psi}{dt} \quad \text{and} \quad \frac{W}{dt} = d \int \frac{\psi^* \psi}{dt} \, d\tau$$

While performing the differentiation of the products in these expressions we can use the time-dependent Schrödinger equation and its complex conjugate analogue to substitute for the differentials $\frac{d\psi}{dt}$ and $\frac{d\psi^*}{dt}$:

$$\frac{dw}{dt} = \psi^* \frac{d\psi}{dt} + \frac{\psi^*}{dt} \psi = \frac{1}{i\hbar} [\psi^* \hat{H}\psi - (\hat{H}\psi)^* \psi]$$

$$\frac{dW}{dt} = \frac{1}{i\hbar} \left[\int \psi^* \hat{H}\psi \, d\tau - \int (\hat{H}\psi)^* \psi \, d\tau \right]$$

We now request that the total probability should not vary in time; that is, that no particles should be 'created' or disappear, which in mathematical terms means $\frac{dW}{dt} = \emptyset$. In this case, the expression inside the brackets must be zero, as $i\hbar \neq \emptyset$

$$\left[\int \psi^* \hat{H}\psi \, d\tau - \int (\hat{H}\psi)^* \psi \, d\tau \right] = \emptyset$$

$$\text{or:} \quad \int \psi^* \hat{H}\psi \, d\tau = \int (\hat{H}\psi)^* \psi \, d\tau$$

and, written in vector notation,

$$\langle \hat{H}\psi | \psi \rangle = \langle \psi | \hat{H}\psi \rangle$$

Operators which fulfil this condition are called *Hermitian* operators; they operate in bra space the same as in ket space. If in addition to hermicity these operators also have the same form in both bra and ket space, that is: $\hat{O} = \hat{O}^*$, they are called *self adjugate.*

2.5 Exact Measurability and Eigenvalue Problems

We will now show some of the specific properties of eigenvalue problems. First, we will recall the method by which we usually evaluate the mean value for a series of equal measurements and its standard deviation, which is given by

$$\sigma = \pm\sqrt{\left(\sum x^2 - \sum (x)^2\right)/(n-1)}$$

An analogous expression can be formed, using the expectation value of an operator $\hat{A}$ (corresponding to the sum of all eigenvalues, $\sum a_i$), and its square operator $\hat{A}^2$, and for the sake of a short notation we will introduce $\langle\hat{A}\rangle \equiv \langle\psi|\hat{A}\psi\rangle/\langle\psi|\psi\rangle$, which leads to

$$\sigma = \Delta A = \pm\sqrt{(\langle\hat{A}^2\rangle - \langle\hat{A}\rangle^2)/(n-1)}$$

Now let us assume that we can measure the value of the variable associated to the operator $\hat{A}$ exactly, i.e., that $\Delta A = 0$. In this case

$$\langle\hat{A}^2\rangle = \langle\hat{A}\rangle^2$$

which can be written explicitly as

$$\langle\psi|\hat{A}^2\psi\rangle/\langle\psi|\psi\rangle = \langle\psi|\hat{A}\psi\rangle^2/\langle\psi|\psi\rangle^2$$

If $\hat{A}$ is a Hermitian operator, we can resolve $\hat{A}^2$ as $\hat{A}\cdot\hat{A}$, take one of the two components to the bra space, and multiply the equation by $\langle\psi|\psi\rangle^2$, obtaining:

$$\langle\hat{A}\psi|\hat{A}\psi\rangle \cdot \langle\psi|\psi\rangle = \langle\psi|\hat{A}\psi\rangle^2$$

This is a special case of equality in the general Cauchy–Schwarz inequality, which also serves to determine the angle θ between two vectors:

$$\langle u|v\rangle^2 \leq \langle u|u\rangle \cdot \langle v|v\rangle$$

and

$$\cos\theta = \langle u|v\rangle/\sqrt{\langle u|u\rangle \cdot \langle v|v\rangle}$$

Equality is only given, if the angle between the two vectors is $\emptyset$ degrees; that is, if the vectors $|u\rangle$ and $|v\rangle$ have the same direction. Therefore, we can conclude that an eigenvalue equation, where the image vectors result by multiplication of the original vectors by scalar quantities, is associated with the exact measurability of the variable belonging to this operator. In Chapter 3 we will see that this finding is the starting point for another fundamental difference between classical and quantum physics, which is essential for an understanding of the interactions between particles and the chemical bond.

2.6 Eigenvalue Problem of Hermitian Operators

If we integrate the eigenvalue equation $\hat{A}|\psi\rangle = a|\psi\rangle$ by multiplying it with $\langle\psi|$ we obtain

$$\langle\psi|\hat{A}\psi\rangle = a\langle\psi|\psi\rangle$$

and we can write the complex conjugate of this expression as

$$\langle\psi|\hat{A}\psi\rangle^* = a^*\langle\psi|\psi\rangle^*$$

With $\langle\psi|\psi\rangle^* = \langle\psi|\psi\rangle$ and assuming $\hat{A}$ a Hermitian operator $\langle\psi|\hat{A}\psi\rangle^* = \langle\hat{A}\psi|\psi\rangle = \langle\psi|\hat{A}\psi\rangle$ we can rewrite the complex conjugate expression as

$$\langle\psi|\hat{A}\psi\rangle = a^*\langle\psi|\psi\rangle$$

from which it becomes clear that $a^* = a$. This means that Hermitian operators always have *real eigenvalues*. This is quite essential in the context of evaluating physical quantities, where imaginary solutions would pose a serious problem of interpretation.

If we now consider two different eigenvectors belonging to different eigenvalues, namely

$$\hat{A}|\psi_1\rangle = a_1|\psi_1\rangle$$

and

$$\hat{A}|\psi_2\rangle = a_2|\psi_2\rangle \quad a_1 \neq a_2$$

and multiply the first equation by $\langle\psi_2|$ and the second one by $\langle\psi_1|$ we obtain

$$\langle\psi_2|\hat{A}\psi_1\rangle = a_1\langle\psi_2|\psi_1\rangle \tag{8}$$

$$\langle\psi_1|\hat{A}\psi_2\rangle = a_2\langle\psi_1|\psi_2\rangle \tag{9}$$

The complex conjugate expression of the latter is

$$\langle\hat{A}\psi_2|\psi_1\rangle = a_2^*\langle\psi_2|\psi_1\rangle$$

If the operator is Hermitian, $\langle\hat{A}\psi_2|\psi_1\rangle = \langle\psi_2|\hat{A}\psi_1\rangle$, and due to the previous statement concerning real eigenvalues, $a_2^* = a_2$. We can thus subtract the complex conjugate of Eq. 9 from Eq. 8 and obtain

$$(a_1 - a_2)\langle\psi_2|\psi_1\rangle = \emptyset.$$

This proves that *eigenvectors* of Hermitian operators belonging to different eigenvalues ($a_1 \neq a_2$) are *orthogonal/unitary*.

We can conclude from the specific properties of the eigenvalue problem of Hermitian operators that the representation matrix of any Hermitian operator will also be Hermitian, and can thus be brought to diagonal form by unitary transformation. The elements of this diagonalised matrix correspond to the eigenvalues, and will all be real.

2.7
The Eigenvalue Equation of the Hamiltonian

The solution of the time-independent Schrödinger equation $\hat{H}|\psi\rangle = E|\psi\rangle$ is the most crucial task in quantum chemistry. An exact solution of it has been achieved so far only for one-electron atoms such as H or He^{+}. In the subsequent chapters we will see how approximate solutions for n-electron systems are achieved. At this point it seems useful, however, to show the general way to deal with eigenvalue problems in practice, and this will be done with the example of the Schrödinger equation. If we consider not just one, but the full set of eigenvalues for $\hat{H}|\psi\rangle = E|\psi\rangle$, we obtain the matrix form

$$\hat{H}\boldsymbol{\Psi} = \mathbb{E}\boldsymbol{\Psi}.$$

If we construct the eigenfunctions in $\boldsymbol{\Psi}$ as linear combinations of basis functions of a finite basis $\{\varphi_i\}$, for the whole set of $|\psi\rangle$ we get $\boldsymbol{\Psi} = \boldsymbol{\Phi}\cdot\mathbb{C}$, where $\mathbb{C}$ is the matrix of the linear combination coefficients. Substituting $\boldsymbol{\Psi}$ in the previous equation by this expression we obtain

$$\hat{H}\boldsymbol{\Phi}\mathbb{C} = \boldsymbol{\Phi}\mathbb{C}\mathbb{E}.$$

Integration of this expression (= scalar multiplication by the corresponding bra vectors) yields

$$\langle\boldsymbol{\Phi}|\hat{H}\boldsymbol{\Phi}\rangle\mathbb{C} = \langle\boldsymbol{\Phi}|\boldsymbol{\Phi}\rangle\mathbb{C}\mathbb{E} = \mathbb{H}^{\varphi}\mathbb{C} = \mathbb{S}\mathbb{C}\mathbb{E}$$

using the abbreviation $\mathbb{H}^{\varphi}$ for the representation of the Hamiltonian in the basis $\{\varphi_i\}$ and $\mathbb{S}$ as the common notation for the overlap integral matrix. This general form of the eigenvalue problem is not suitable for a numerical solution due to the overlap integral matrix on the right-hand side of the equation. However, we know that by a Löwdin orthonormalisation we can easily transform any basis $\{\varphi_i\}$ to an orthonormal basis $\{\varphi_i'\}$ according to $\boldsymbol{\Phi}' = \boldsymbol{\Phi}\mathbb{S}^{-\frac{1}{2}}$. In our linear combination of basis functions, this corresponds to a transformation of the coefficient matrix $\mathbb{C}$, according to $\mathbb{C}' = \mathbb{S}^{\frac{1}{2}}\mathbb{C}$. This can be easily shown by comparing

$$\boldsymbol{\Psi} = \boldsymbol{\Phi}\mathbb{C} \quad \boldsymbol{\Psi} = \boldsymbol{\Phi}'\mathbb{C}' = \boldsymbol{\Phi}\mathbb{S}^{-\frac{1}{2}}\mathbb{C}'$$

from which it follows that $\mathbb{C} = \mathbb{S}^{-\frac{1}{2}}\mathbb{C}'$ and thus $\mathbb{C}' = \mathbb{S}^{\frac{1}{2}}\mathbb{C}$.

Having performed the Löwdin transformation, which reduces the overlap matrix to the unit matrix, we thus have the eigenvalue problem given as

$$\mathbb{H}^{\varphi'}\mathbb{C}' = \mathbb{C}'\mathbb{E} \quad \text{or} \quad (\mathbb{H}^{\varphi'} - \mathbb{E})\mathbb{C}' = \emptyset.$$

This actually corresponds to a system of n linear equations, which – in analogy to equations common in astronomy – are called '*secular equations*' and the corresponding determinant $|\mathbb{H}^{\varphi'} - \mathbb{E}|$ is the '*secular determinant*'.

We know that the matrix $\mathbb{H}^{\varphi'}$ is Hermitian, and can thus be brought to diagonal form $\mathbb{E}$ by unitary transformation:

$$\mathbb{H}^{\varphi'}\mathbb{C}' = \mathbb{C}'\mathbb{E} \quad \mathbb{C}'^{-1}\mathbb{H}^{\varphi'}\mathbb{C}' = \mathbb{E}.$$

This transformation is actually achieved with the eigenvector matrix $\mathbb{C}$ and its inverse, and such a unitary transformation by the eigenvector matrix is called *principal component transformation.*

In practice this *diagonalisation* process is achieved numerically. The most illustrative procedure (Jacobi rotation) is a sequential annihilation of non-zero off-diagonal values, starting with the largest one, by multiplying the matrix of the operator with transformation matrices and continuing until no off-diagonal element is larger than a limit, mostly set to 10^{-12}. From the set of all transformation matrices used one obtains the eigenvector matrix, and the resulting diagonal matrix contains the eigenvalues. In fact this procedure is a stepwise principal component transformation. For the performance of diagonalisation in a computer, there are many more efficient programs available nowadays, which are either built-in in all common quantum chemical program packages or are available separately and thus of no immediate concern to the chemist. Rather, the chemist's task is to find an appropriate representation matrix of the Hamiltonian.

2.8 Eigenvalue Spectrum

An operator can have numerous different eigenvalues, associated with different eigenvectors, thus defining a set of discrete values that the associated variable can assume and their corresponding eigenstates. This is quite a different situation from that in classical physics, where it is assumed that every physical quantity can vary continuously, thus allowing an infinite number of states. When we recall the initial problems of the Bohr–Sommerfeld atomic model, in which one had to assume special stable orbits for the electrons without knowing a physical reason for this stability, we see that the quantum mechanical treatment leading to eigenvalue problems of operators provides a simple solution for this dilemma. Only a finite number of states exists, for which the value of a physical observable is well defined and exactly measurable. It is possible, however, that there is more than one state associated with the same eigenvalue, and such an eigenvalue is called *degenerate*. Chemists know implicitly this 'degeneration' from their classification of the electrons in the periodic system of the elements, where p electrons are 3-fold degenerate (p_x, p_y, p_z electrons have the same energy), and d and f electrons are 5- and 7-fold degenerate, respectively.

The full eigenvalue spectrum of an operator not only covers the aforementioned *discrete* spectrum, but also contains another, *continuous* part. 'Bound' states, where a stabilisation of the particles is given, correspond to the discrete part of the eigenvalue spectrum. Thus, any atom or molecule – even in an excited or partly ionised state – is characterised by a discrete eigenvalue spectrum. 'Nonbound' states (e.g., a free electron scattered by an atom) are described by the continuous part of the operator's spectrum, and indeed the electron in our example can assume any arbitrary energy value. The discrete part of the eigenvalue spectrum can be treated in Hilbert space, and as chemists are mainly interested in bound states of matter, Hilbert space is mostly sufficient for vector space theory of matter. It should be mentioned that also the continuous part can be transferred to Hilbert space, not in terms of eigenvalues but via eigendifferentials. This will not be needed, however, for all further considerations in this book, as we will only deal with bound states.

Test Questions Related to this Chapter

1. What interpretation do we give to the wave equation?
2. Why cannot particles be treated as waves propagating in a medium or as wave packets?
3. What are the two fundamental postulates of quantum mechanics, and in what aspect are they in contradiction to classical physics?
4. Which basic input do we need to derive the time-dependent Schrödinger equation?
5. How can we derive the time-independent Schrödinger equation?
6. What is the fundamental assumption needed to show that the Hamiltonian is a Hermitian operator?
7. If an operator fulfils an eigenvalue equation, what does it imply for the associated variable?
8. Which parts of the eigenvalue spectrum do we know, and to what states of a system are they related?
9. What are the properties of a representation matrix of a Hermitian operator and of its eigenvalues and eigenvectors?
10. Why must we perform a Löwdin transformation before solving the eigenvalue problem of the Hamiltonian?

3
Consequences of Quantum Mechanics

Even if the number of atoms is infinite, the number of their different shapes is limited.

Heraklitos, 6th century BC

The findings achieved in Chapter 2 by introducing the probability concept and defining a vector space describing matter in bound states under the conditions of quantisation lead to further important and far-reaching conclusions on the nature of matter, which will be outlined in this chapter.

3.1
Geometrical Interpretation of Eigenvalue Equations in Vector Space

As stated above, in an eigenvalue equation a probability vector is mapped by an operator onto itself, multiplied by scalar values corresponding to the eigenvalues of the associated variable. Geometrically this means simply an 'elongation' or 'compression' of the probability vector which, however, maintains its 'direction' in the vector space. Additionally, this behaviour was recognised as a condition for the exact measurability of the physical observable. One could imagine, however, that another operator acting on this vector would not only change its length, but also its direction.

This can be illustrated by the eigenvalue equations of two different operators

$$\hat{A}|\psi_a\rangle = a|\psi_a\rangle \qquad \hat{B}|\psi_b\rangle = b|\psi_b\rangle$$

we could now find $\hat{B}|\psi_a\rangle = b|\psi_a\rangle$, but alternatively we could also find:

$$\hat{B}|\psi_a\rangle \neq b|\psi_a\rangle$$

which means that in this case $|\psi_a\rangle$ is not an eigenvector of the operator $\hat{B}$ or, in other words, if the system is in a state characterised by $|\psi_a\rangle$, the variable associated with the operator $\hat{B}$ cannot be exactly determined. This is another principally new finding in physics: in classical physics it was believed that all observable

The Basics of Theoretical and Computational Chemistry. Edited by B. M. Rode, T. S. Hofer and M. D. Kugler

ISBN: 978-3-527-31773-8

quantities in a system could be determined simultaneously with arbitrary accuracy, if suitable instruments were available. Quantum physics shows, however, that an inherent incompatibility of observables could exist preventing their accurate simultaneous determination. According to this, two physical observables are classified as *compatible* with each other, if they can both be accurately determined at the same time, and as *complementary*, if they cannot. In the following we will introduce a convenient instrument to distinguish between these two cases.

3.2 Commutators and Uncertainty Relationships

In general, operators – and thus also their representation matrices – are not commutative, and in order to check the difference between $\hat{A}\hat{B}$ and $\hat{B}\hat{A}$ an operator is defined, the so-called *commutator*.

$$[\hat{A}, \hat{B}] = \hat{A}\hat{B} - \hat{B}\hat{A}$$

Two commuting operators can be characterised as following:

$$\hat{A}|\psi_a\rangle = a|\psi_a\rangle \qquad \hat{B}|\psi_b\rangle = b|\psi_b\rangle$$

and if $|\psi_a\rangle = |\psi_b\rangle \equiv |\psi\rangle$, the operation of $[\hat{A}, \hat{B}]$ on $|\psi\rangle$ produces

$$\begin{aligned}[\hat{A}, \hat{B}]|\psi\rangle &= \hat{A}b|\psi\rangle - \hat{B}a|\psi\rangle \\ &= ab|\psi\rangle - ba|\psi\rangle = (ab - ba)|\psi\rangle = |\emptyset\rangle\end{aligned}$$

Thus, if $|\psi\rangle$ is an eigenvector for both $\hat{A}$ und $\hat{B}$, which means that both variables are compatible, the commutator becomes an annihilation operator or zero operator:

$$[\hat{A}, \hat{B}] = \hat{\emptyset} \quad \textit{compatibility}$$

On the other hand, if the commutator does not act as zero operator, the operators $\hat{A}$ and $\hat{B}$ do not have common eigenvectors and their associated observables are not exactly determined at the same time:

$$[\hat{A}, \hat{B}] \neq \hat{\emptyset} \quad \textit{complementarity}.$$

We will now apply this knowledge about the possibility of recognising compatibility/complementarity of observables with the help of commutators to some important operators. The first example deals with the operators of energy and time acting on an arbitrary function f(t):

$$\hat{t} = t; \quad \hat{E} = i\hbar\frac{\partial}{\partial t}$$

$$\begin{aligned}
[\hat{t}, \hat{E}]f(t) &= \left[t, i\hbar\frac{\partial}{\partial t}\right]f(t) \\
&= t \cdot i\hbar\frac{\partial}{\partial t}f(t) - i\hbar\frac{\partial}{\partial t}(t \cdot f(t)) \\
&= t \cdot i\hbar\frac{\partial}{\partial t}f(t) - t \cdot i\hbar\frac{\partial}{\partial t}f(t) - i\hbar f(t) \\
&\Rightarrow [\hat{t}, \hat{E}] = -i\hbar \neq \hat{\emptyset}
\end{aligned}$$

where $\hat{t}$ and $\hat{E}$ are thus not commuting and the associated observables *time and energy are complementary* – that is, they cannot be exactly determined simultaneously. This leads us to the first of a series of relationships called *Heisenberg uncertainty relationships*:

$$\Delta E \cdot \Delta t \geq const.$$

The fundamental meaning of this relationship is that energy is not exactly determined for a small time interval Δt. We will soon see that this relationship is of utmost importance for understanding the action of forces between particles.

Our second example deals with the operators of space and momentum. For the one-dimensional case this means analysing the commutator between $\hat{x}$ and $\widehat{p_x}$:

$$\begin{aligned}
[\hat{x}, \widehat{p_x}]f(x) &= x\left(-i\hbar\frac{\partial}{\partial x}|i\rangle\right)f(x) + i\hbar\frac{\partial}{\partial x}|i\rangle(x \cdot f(x)) \\
&= x\left(-i\hbar\frac{\partial}{\partial x}|i\rangle\right)f(x) + xi\hbar\frac{\partial}{\partial x}|i\rangle f(x) + i\hbar f(x) \\
&\Rightarrow [\hat{x}, \widehat{p_x}] = i\hbar \neq \hat{\emptyset}.
\end{aligned}$$

This leads to another fundamental uncertainty relationship, namely the *space-momentum uncertainty*, which will play an important role in the explanation of the chemical bond:

$$\Delta x \cdot \Delta p_x \geq const. \quad \text{or} \quad \Delta r \cdot \Delta p \geq const.$$

for location in space and the total momentum. This relationship will immediately prove helpful in answering a simple question, arising when one thinks of the hydrogen atom in electrostatic terms: the positively charged proton and the negatively charged electron should attract each other, until they fuse to a neutron. But why does this not happen? The answer results from the uncertainty relationship, in that: the closer the electron comes to the nucleus, the smaller its uncertainty in space Δr becomes, and the larger Δp becomes. Thus, the average momentum and the kinetic energy of the electron prevents it from being 'absorbed' by the nucleus (it should be noted, however, that under the influence of other than electromagnetic forces this fusion can take place, e.g., in a radioactive processes or under the influence of gravitation in the formation of neutron stars).

In Chapter 4 we will encounter several other important consequences of the uncertainty relationships for chemistry. Before that, we will deal with the relevance of the time–energy relationship for the understanding of forces in nature.

3.3 Virtual Particles and the Forces in Nature

To date, physics knows four types of forces acting in nature:

1. Electromagnetic forces
2. Weak nuclear forces
3. Strong nuclear forces
4. Gravitation

Since the early 20th century until the present day, the unification of these forces into one common theoretical framework has been the most challenging task of physics, and to date this has been achieved only for the first three forces listed. The energy–time uncertainty relationship proved essential not only for this purpose, but also for the understanding of interactions between particles.

Due to the energy–mass relationship $E = mc^2$, we can rewrite the energy–time uncertainty relationship as

$$\Delta E \cdot \Delta t \geq const. \quad \Delta m \cdot \Delta t \geq const.$$

This means that, for a given time interval Δt, a particle with the mass Δm can be 'created' from 'nothing'. The smaller Δm, the larger is the lifetime of this particle, and the further it can travel in space. Such particles are called *virtual particles*, and they are considered as the 'force-transmitting' particles responsible for interactions between 'matter' particles.

For the electromagnetic force, the force-transmitting particles are (virtual) photons. We can understand now, how two electrons in a vacuum can interact with each other without any medium connecting them in order to propagate the repulsive force between them: each electron emits virtual photons, which 'hit' the other electron. The larger the distance between the electrons, the weaker this interaction becomes, as fewer virtual photons will reach the other electron (the phenomenological description of this behaviour is Coulomb's law). However, due to the extremely small mass Δm of the photons, Δt, the reach of the virtual photons is quite large in terms of atomic dimensions. This is quite different in the case of the nuclear forces, which are transmitted by the so-called W and Z particles (weak forces) and gluons (strong forces), as these particles have much larger masses than photons. Consequently, the nuclear forces are restricted to a much smaller dimension, namely to the atomic nucleus and its immediate environment.

Gravitation is the weakest and farthest-reaching force in nature, and the search for force-transmitting 'gravitons' is continuing. In theoretical research on gravitation, the combination of the classical gravitation force with quantum theory is also a main topic of contemporary studies.

The first unification of forces was achieved for electromagnetic and weak nuclear forces, leading to the 'electroweak force'. This can be realised experimentally at energies of $\sim$100 GeV (also called the 'unification temperature' of the electroweak force). Further unification with strong nuclear forces requires an energy of $\sim 10^{13}$ GeV, and the construction of increasingly powerful particle accelerators aims at this target. In physics, this unification is termed *GUT (Grand Unified Theory)*. This theory provides a good explanation for the consecutive separation of forces after the 'Big Bang' at the start of our universe. However, further unification with gravitation encounters a serious problem: the theoretical energy needed for unification (with the classical gravitation force) is $\sim 10^{19}$ GeV, which already corresponds to a mass forming a 'black hole'. This is the reason why new theoretical approaches to gravitation appear unavoidable, and attempts to achieve a 'Unified Field Theory' and to combine the (classical) theory of relativity and quantum theory are a prominent topic in theoretical physics research.

Test Questions Related to this Chapter

1. How can we characterise the angle between eigenvectors that belong to different operators, but do not produce eigenvalues for both operators?
2. How can we find out whether two operators have the same eigenvectors?
3. Which forces in nature could be unified in theory, and which not?
4. Why are electromagnetic forces much more far-reaching than nuclear forces?
5. How can we explain the repulsion of two protons *in vacuo*?
6. What is the meaning of non-commuting operators with respect to their associated variables?
7. By which particles are the weak nuclear forces transferred?

4
Chemistry and Quantum Mechanics

You should proceed via falsifications, and only if you can exclude everything else, attempt an affirmative statement.

Francis Bacon (Baron Verulam), 17th century.

This chapter deals with the specific consequences of quantum mechanics for chemical systems, and will investigate some of the common models developed on the basis of early quantum mechanical studies. Such investigations are still widely used in many chemical textbooks, even though they are sometimes questionable – if not wrong – from a rigorous quantum theoretical viewpoint. We will also encounter another fundamental concept – the Pauli antisymmetry principle – and introduce spin as a non-classical variable. This will lead us finally to a contemporary theory of chemical bonding.

4.1
Eigenvalue Problem of Angular Momentum and 'Orbital' Concept

In the early stage of quantum theory, *Bohr and Sommerfeld* developed their famous atomic model, in which electrons were assigned to orbits of discrete energies, claimed as 'stable' – that is, where the electrons would not lose energy despite being a moving charged particle. These orbits were characterised by the *quantum numbers* n, l, and m, called principal, azimuthal, and magnetic. The first number defined the level of energy, the second, and the third one the 'type' and the 'shape' of the orbit. These numbers were then used to classify the electronic structure of the elements in the Periodic Table, with n indicating the period, l the type of electrons ($l = 0, 1, 2, 3$ corresponding to 's, p, d, f'-electrons), and m to distinguish the degenerate states of these electrons, e.g., p_x, p_y, p_z. When the Schrödinger equation for the hydrogen atom was solved, these quantum numbers resulted as a part of the solutions, with n being associated with the radial part of the solutions, and l and m with the angular part (the solution was achieved in polar coordinates). From this, the concept of 'orbitals' for the electrons was developed, and the s, p, d, and f-electrons used in the Periodic Table of the elements were all assigned the well-known shapes found in all chemical text-

The Basics of Theoretical and Computational Chemistry. Edited by B. M. Rode, T. S. Hofer and M. D. Kugler

ISBN: 978-3-527-31773-8

books. We should not forget, however, that all of these model pictures are based on the solution for a one-electron system.

Thus, we will now examine the correctness of this model picture in the light of our knowledge of complementary and compatible variables. If a chemical system, atom or molecule, is in a defined state of energy, it is in an eigenstate of the Hamiltonian. The quantum numbers l are the eigenvalues of the operator of the amount of the angular momentum $\hat{L}^2$, and the commutator of both operators, $[\hat{H}, \hat{L}^2]$ will tell us whether energy and l are compatible variables. $\hat{H} = \hat{T} + \hat{V}$, and we find:

$$[\hat{T}, \hat{L}^2] = \hat{\emptyset}$$

but for the potential energy operator we find

$$[\hat{V}, \hat{L}^2] = \hat{\emptyset}$$

only, if the potential is spherically symmetric. Therefore, in all other cases, where we cannot assume a spherically symmetric 'field',

$$[\hat{H}, \hat{L}^2] \neq \hat{\emptyset}$$

Consequently, the values of l are not exactly determined. For chemical systems this means that the assumption of well-defined s, p, d, f-electrons might be a good approximation for isolated atoms, for which we can assume a rather spherical symmetry of the electron density distribution, but never for molecules. Therefore, the eigenfunctions of $\hat{L}^2$, which are the so-called *'spherical harmonics'* $Y_{\theta,\varphi}^{l,m}$, classifying the aforementioned shapes of the orbitals in terms of the polar coordinates θ and φ as a function of the quantum numbers l and m, are also not well defined. This makes it clear that, in terms of strict theory, one cannot assume s, p, d, f-electrons to exist in a molecule, and all models built on this hypothesis must therefore be questioned. Hence, this is an appropriate point briefly to discuss the two main model approaches used to describe molecules on the basis of the orbital concept.

4.2 Molecular Orbital and Valence Bond Models

Both the MO and VB models start from the solutions found for the Schrödinger equation of the hydrogen atom, and assign corresponding *atomic orbitals* to each electron of the individual atoms making up the chemical system. In the *MO model*, these orbitals are combined by linear combination to new functions, the *'molecular orbitals'*, usually spread over the whole system, be it a molecule or a set of molecules. These new functions are not very perspicuous, especially for larger molecules, and the construction of the molecular orbitals is in most cases a numerical task for the computer.

In the *VB model*, the atomic orbitals retain their original shape, and one looks only for the overlap of the orbitals of different atoms, defining these overlap regions as chemical bonds. This concept is more illustrative, but it soon leads to problems of how to explain the known structure of simple molecules, (e.g., the tetrahedral structure of methane). An additional hypothesis attempted to solve this problem, postulating the 'promotion' of one s-electron of carbon to a p-state, followed by a '*hybridisation*' to a so-called sp^3 'state' with four equivalent atomic orbitals. Although this concept of 'hybridisation' by no means reflects a physical property of any system, it has been vividly extended to other 'states' (sp^2, sp), not only for carbon, but also for other atoms, and has become a part of the (mostly organic) chemist's everyday language. Often, it is even mixed up with the MO model and terminology, despite being a part of the VB approach.

Both approaches can be evolved to quantitative formalisms, and when they are executed at the highest level of accuracy, they become equivalent. The qualitative orbital models found in many textbooks and publications are usually a rather obscure and physically meaningless mixture of both approaches – additionally rooted on the one-electron simplification – and it can only be hoped that chemical language will one day be freed from this burden, which is in contradiction to rigorous theory and does not provide any realistic explanations in chemistry. Sometimes these models seem to be helpful to 'rationalise' the description of certain subsets of experimental observations. Orbital symmetry rules and the HOMO-LUMO concept (*vide infra*) could be cited as examples of such putatively 'helpful' tools, but everybody knows that these concepts usually require an ever-increasing number of exceptions to prove (?) the rules.

One very curious (albeit almost artistic) model in chemistry is the 'backbonding concept', which is mainly used in metal–organic complex chemistry. The 'art' of this concept consists of a concentrated violation of physical principles and a thorough mix of the MO and VB models: electrons are treated as if they were distinguishable and had sharp fixed energy levels, irrespective of the occupation of associated functions. The central metal provides 'd-electrons' and 'empty d-orbitals' (VB!), while the ligands provide electrons in 'π-orbitals' (MO!) plus some empty 'π-orbitals', and the (apparently distinguishable) electrons of one partner migrate to the 'orbitals' of the other one, to lend special stability to the compound.... Hopefully, this model will – together with other similar constructions – soon end up in the storage place of the phlogiston theory of redox processes.

The ligand field model (also known as the crystal field model) is another very common simple concept in complex chemistry and, due to its simplicity, is much more satisfactory in terms of physics. Here, the ligands are treated as point charges or dipoles in a geometrical arrangement reflecting the structure of the complex, thus exerting a perturbation on the electron distribution by a structured electrostatic field. The electronic description of the central metal is maintained at the level of the periodic system, in the form of atomic orbitals assigned to each electron, which seems well acceptable assuming spherical symmetry for the metal atom. Due to the electrostatic perturbation by the ligands, symmetry is lowered to octahedral, tetrahedral or even lower grade, and thus the degeneration of the orbitals (mostly d orbitals are considered) is broken up. The differences between the energy values of the newly positioned orbitals and electronic transitions between them are used to describe the absorption of the complexes in the UV/VIS region, and to discuss the Jahn–Teller effect, predicting that a descent to a lower than octahedral or tetrahedral symmetry of a metal complex leads to a lower electronic energy for d^1 or d^9 metal ions. The ligand field model implies that the interaction between central metal and ligands should be mainly of electrostatic nature, and thus it becomes worse if covalent bond contributions increase. Within all these limitations, however, the ligand field model appears to be a useful qualitative tool.

Apparently, only the quantitative treatment of a large number of systems will make it possible to find more general qualitative rationalisation concepts in chemistry based on quantum theory, and at a later stage we will see that in this way we can prove other model concepts to be closer to modern theory than the orbital pictures of the unfortunately widespread qualitative MO and VB models. A recommendation how to deal with the majority of these qualitative models is given in Fig. 4.1.

Fig. 4.1 Recommended procedure for the treatment of most qualitative MO and VB models in chemistry.

4.3 Spin and the Antisymmetry Principle

When angular momentum in quantum mechanics of atoms was analysed in more detail it was discovered that in quantum systems the conservation of momentum could not be achieved with classical angular momentum alone, but that an additional component was required to fulfil this fundamental principle. The relativistic treatment of the Schrödinger equation by *Dirac* led to the same result, and consequently an additional component of angular momentum, named *spin*, had to be introduced. The total angular momentum of an electron thus results from the combination of both components as $|j\rangle = |l\rangle + |s\rangle$. The eigenvalue associated with the spin variable, for which no classical analogue exists, is also known as the fourth quantum number and it was found that in terms of the Bohr–Sommerfeld atomic model two electrons in the same system could not have four identical quantum numbers. This is the older formulation of the *Pauli Exclusion Principle.*

When we consider a system of n elementary particles and describe each of these particles by an individual one-particle function and the total system as a normalised product of all one-electron functions

$$|\Psi\rangle = \frac{1}{\sqrt{n!}}[\psi_1(1) \cdot \psi_2(2) \cdot \ldots \cdot \psi_i(i) \cdot \psi_j(j) \cdot \ldots \cdot \psi_n(n)]$$

we could exchange the coordinates of two electrons i and j according to

$$\widehat{P_{ij}}|\Psi\rangle = \frac{1}{\sqrt{n!}}[\psi_1(1) \cdot \psi_2(2) \cdot \ldots \cdot \psi_i(j) \cdot \psi_j(i) \cdot \ldots \cdot \psi_n(n)] = |\Psi'\rangle$$

with $\widehat{P_{ij}}$ being a permutation operator. As electrons are indistinguishable, this should not change any of the physical properties of the system. The function $|\Psi\rangle$ itself does not have any physical meaning, only the product $\langle\Psi|\Psi\rangle$, or for real functions Ψ^2, is associated with probability and thus of physical significance. Thus, two cases could result from the action of the permutation operator $\widehat{P_{ij}}$, namely

$$\widehat{P_{ij}}|\Psi\rangle = \pm|\Psi\rangle,$$

and indeed we find both cases in nature, the symmetric $(+)$ and the antisymmetric $(-)$ behaviour of the total function $|\Psi\rangle$. This leads to a fundamental classification of elementary particles: those with *antisymmetric* behaviour are called *Fermions*, and those with *symmetric* behaviour *Bosons*, according to the Fermi–Dirac and Bose–Einstein statistics that they obey.

Electrons are Fermions, and any function describing a system of electrons must, therefore, be antisymmetric. This is also known as the *Pauli Antisymmetry Principle.*

Moreover, symmetric/antisymmetric behaviour serves as a classification of all elementary particles, and it is also associated with spin properties. All *matter particles* such as electrons and protons are Fermions, and their spin value is $\frac{1}{2}$ or an uneven multiple of it. On the other hand, all *energy-transmitting particles* such as photons and gluons are Bosons, and their spin value is an integer.

We have mentioned that the total angular momentum results as a combination of the 'orbit' angular momentum and the spin, according to $|j\rangle = |l\rangle + |s\rangle$. For a system with n electrons, this combination of the contributions of each electron, called *Spin-Orbit Coupling*, can occur according to two different schemes. For lighter atoms, the so-called *Russell–Sounders Coupling* dominates:

$$J = L + S \quad \text{with } L = \sum_i l_i \text{ and } S = \sum_i s_i$$

For heavier atoms, the so-called *j-j-Coupling* prevails:

$$J = \sum_i (l_i + s_i).$$

Spin-orbit coupling is another aspect to be considered, when the justification of sharp eigenvalues for l is discussed, as it means that L is no more a 'good' quantum number (another common expression for an exactly determinable value) in an atom, but only J. In molecules even J is not a 'good' quantum number, due to the lack of spherical symmetry, as mentioned above. Consequently, the classification of electronic states of the elements in the periodic system by s, p, d, f-electrons is also only valid if we neglect spin-orbit coupling. Otherwise we must introduce a modified classification scheme based on the quantum number J and the different values it can assume for a given element in different electronic configurations. Notably, this is done when discussing electronic spectra of the elements in X-ray fluorescence spectroscopy.

4.4 The Virial Theorem

Before we start to investigate the reasons why atoms bind to each other to form molecules, we must deal with a general theorem referring to the relationship between kinetic and potential energy in the equilibrium state of a chemical system. This theorem will not only prove essential for discussing the chemical bond, but also supply a convenient measure for the quality of quantitative molecular orbital calculations.

Let us assume a diatomic molecule, which is given by a probability function $|\Psi(r_i)\rangle$, where the space coordinates r_i will be used for a variational ansatz with $r_i' = \eta r_i$. The normalised function $|\Psi_\eta\rangle$ has the following form:

$$\psi_\eta = \eta^{\frac{3n}{2}}\psi(\eta r_1, \eta r_2, \ldots, \eta r_n)$$

The expectation value of the total energy is then given by

$$E = \langle \psi_\eta | \hat{T}\psi_\eta \rangle + \langle \psi_\eta | \hat{V}\psi_\eta \rangle$$

(the denominator $\langle \psi_\eta | \psi_\eta \rangle$ has disappeared due to the normalisation of the function).

We also vary R, the distance between the two nuclei, as $\rho = \eta R$. Then, we can write the contributions for potential and kinetic energy as follows:

$$\langle \hat{V} \rangle = \eta \langle \psi | \hat{V}\psi \rangle_\rho$$
$$\langle \hat{T} \rangle = \eta^2 \langle \psi | \hat{T}\psi \rangle_\rho$$

where both expressions now depend on ρ as variable. The coefficients η and η^2 must now be added, as $\hat{V}$ and $\hat{T}$ contain the terms $\frac{1}{R}$ and $\frac{1}{R^2}$, respectively. The expression for the total energy thus results as

$$E = \eta^2 \langle \hat{T} \rangle_\rho + \eta \langle \hat{V} \rangle_\rho$$

We will now optimise the energy with respect to η, according to $\frac{\partial E}{\partial \eta} \Rightarrow \emptyset$, which leads to the following expression:

$$\frac{\partial E}{\partial \eta} = 2\eta \langle \hat{T} \rangle_\rho + \langle \hat{V} \rangle_\rho + \eta^2 R \frac{\partial \langle \hat{T} \rangle_\rho}{\partial \rho} + \eta R \frac{\partial \langle \hat{V} \rangle_\rho}{\partial \rho} = \emptyset.$$

Let us assume that $|\Psi_\eta\rangle = |\Psi_{exact}\rangle$, i.e. $\frac{\partial E}{\partial \eta} = \emptyset$ for $\eta = 1$. Then we obtain

$$2\langle \hat{T} \rangle_R + \langle \hat{V} \rangle_R + R\frac{\partial E}{\partial R} = \emptyset$$

and with $E = \langle \hat{T} \rangle + \langle \hat{V} \rangle$ we obtain

$$\langle \hat{T} \rangle = -E - R\frac{\partial E}{\partial R} \quad \langle \hat{V} \rangle = 2E + R\frac{\partial E}{\partial R}$$

for kinetic and potential energy. For the equilibrium distance of the two atoms, corresponding to an energy minimum, $\frac{\partial E}{\partial R} = \emptyset$, and the expressions reduce to

$$\langle \hat{T} \rangle = -E \quad \langle \hat{V} \rangle = 2E \quad \langle \hat{V} \rangle = -2\langle \hat{T} \rangle$$

This is the *virial theorem*, which states that in the equilibrium state of a system the potential energy is twice the total energy, and the kinetic energy equals the amount of total energy, but has the opposite sign.

4.5 The Chemical Bond

4.5.1 General Considerations and One-Electron Contributions

Soon after the establishment of the periodic system of the elements and the recognition of the particular stability of the rare gases, the question of chemical bonding was rationalised on the basis of the electronic configuration of the elements and their assumed tendency to achieve a rare gas electron configuration by donating or accepting electrons. This led to the well-known concept distinguishing three types of bonds, namely 'electrostatic', 'covalent', and 'metal' binding, which until now have proven useful for a classification of binding in molecules. In particular, the classification of 'electrostatic' and 'covalent' contributions to or 'character' of bonds is very common and has been related to the amount of overlap (VB) or mixing (MO) of atomic orbitals located at neighbouring atoms.

The terminology of a varying 'character' of a bond already indicates a rather continuous transition between the extremes of a purely 'electrostatic' and an apolar covalent bond, and thus the possibility to describe all of the phenomena in a single theoretical framework. A detailed quantum theoretical analysis of the chemical bond has been performed, based on an earlier study of *Hellmann*, by *Kutzelnigg* during the 1970s, and we will closely follow these investigations here.

We start from the simplest possible molecule, H_2^+, being formed from a hydrogen atom and a proton. For the electron we use a simple probability function, corresponding to an 's-orbital', namely $\psi = N \cdot e^{-\eta r}$. Figure 4.2 shows the kinetic, po-

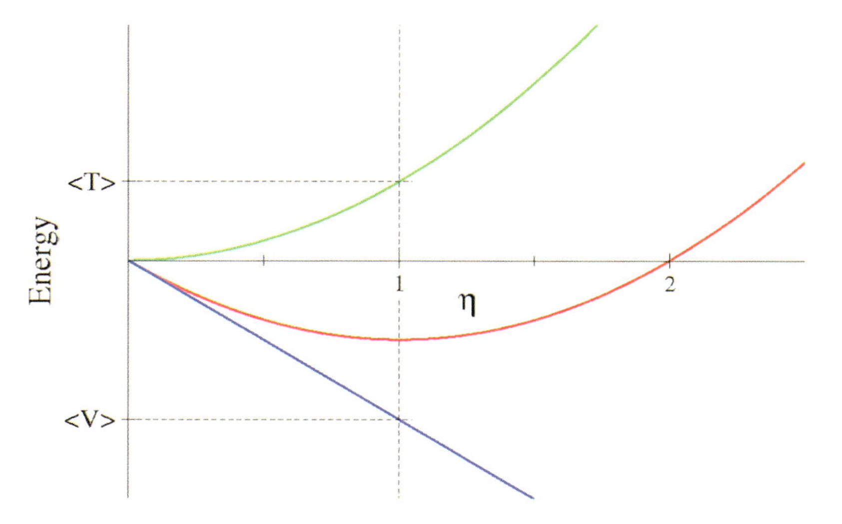

Fig. 4.2 Energy contributions in the H atom as a function of the exponent η of the function for the electron.

tential and total energy of the hydrogen atom as a function of η, and we see that a minimum of total energy is obtained for $\eta = 1$. As $\langle\hat{V}\rangle = -\eta$ and $\langle\hat{T}\rangle = \frac{\eta^2}{2}$, we also see that for $\eta = 1$ the virial theorem is fulfilled: $\langle\hat{V}\rangle = -2\langle\hat{T}\rangle$.

When we move to the H_2^+ molecule we have two nuclei, A and B, to which we assign the same probability function for the electron, writing them in a simplified form as $|a\rangle$ and $|b\rangle$:

$$|\psi_a\rangle \equiv |a\rangle = N.e^{-\eta r_a} \qquad |\psi_b\rangle \equiv |b\rangle = N.e^{-\eta r_b}$$

In these, as in all following expressions, N stands for the normalisation factor. The probability of finding an electron in the surrounding of nucleus A is given by a^2, and in the surrounding of nucleus B by b^2. As the nuclei and the functions are equivalent, the probability for the electron to be located near A or near B is the same. One could consider now simply to form an expression for the electron density ρ with the average of a^2 and b^2, and as this would rather correspond to a quasi-classical treatment (probabilities on the one hand, but classical addition of variables on the other), we will call this a 'quasi-classical' density:

$$\rho_{qc} = \frac{1}{2}(a^2 + b^2)$$

The black curve 1 in Figure 4.3 shows the energy of H_2^+ as a function of the internuclear distance within this quasiclassical approach, which evidently does

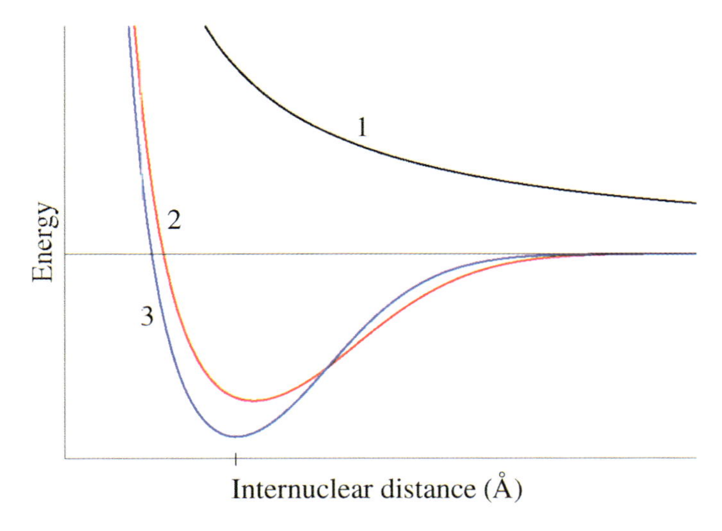

Fig. 4.3 Energy of the H_2^+ molecule as a function of the interatomic distance. black: quasiclassical treatment; red: after superposition of functions; blue: after inclusion of contraction.

not lead to a stable molecule. The only energy contribution that can be treated classically is, therefore, the repulsion between the nuclei.

The mistake of the quasiclassical approach is to neglect the *mutual interference* of the probability functions, forming a new over-all probability function according to

$$|\psi\rangle = N(|a\rangle \pm |b\rangle)$$

From the two possibilities of positive and negative *superposition* we will focus on the positive case ψ_+, for which the overall probability can be calculated by

$$\int (\psi_+)^2 \, d\tau = N^2 \int (a+b)^2 \, d\tau = N^2 \int (a^2 + 2ab + b^2) \, d\tau$$

$$= N^2 \cdot \left[\int a^2 \, d\tau + \int b^2 \, d\tau + \int 2ab \, d\tau \right] = N^2(1 + 1 + 2S)$$

where $S = \int ab \, d\tau$, the overlap integral. We can now normalise the function, which yields

$$\psi_+ = \frac{1}{\sqrt{2(1+S)}} [|a\rangle + |b\rangle]$$

and with this function we obtain a new expression for the electron density, taking into account the superposition of probability functions:

$$\rho_{sup} = \frac{1}{\sqrt{2(1+S)}} [a^2 + b^2 + 2ab]$$

If we form the difference between ρ_{sup} and ρ_{qc}, we obtain

$$\Delta\rho = \rho_{sup} - \rho_{qc} = \frac{1}{\sqrt{2(1+S)}} (2ab - Sa^2 - Sb^2)$$

which means that electron density will be shifted from the regions where only $|a\rangle$ or only $|b\rangle$ have large values (near the nuclei) to the internuclear region, where both of them have comparable values. This is illustrated in Figure 4.4, where the change in electron density distribution is shown: the black curve 1 represents the quasiclassical approach, the red curve 2 the distribution with the superimposed probability functions. The red curve 2 in Figure 4.3 shows the consequences on total energy, and displays a stabilised minimum for the molecule formation. This is striking proof that the mutual superposition of the probability functions, leading to a new, molecular probability function, is an essential factor of stabilisation for the chemical bond.

The reason for this stabilisation is often claimed to be the attraction of the electron by two nuclei instead of one nucleus, which would mean an improved poten-

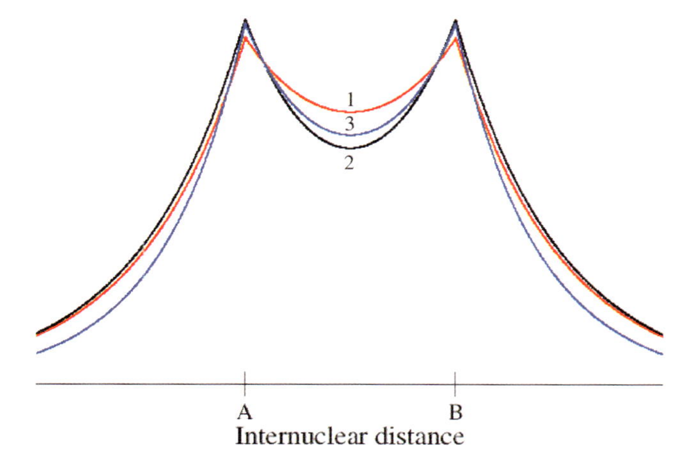

Fig. 4.4 Electron density distribution in the H_2^+ molecule. Black curve, quasiclassical treatment; red curve, after superposition of functions; blue curve, after inclusion of contraction.

tial energy. The counter-argument to this is that the electron density accumulated in the binding region – and thus interacting with both nuclei – had to be removed from the regions near to a nucleus, where potential energy stabilisation is evidently higher. Thus, one must examine the contribution of kinetic energy, and the space-momentum uncertainty relationship $\Delta x \cdot \Delta p_x \geq const$ is the key to this issue. Going from the atom to the molecule, the spatial distribution of electron density along the bond axis (here assumed to be the x-axis) becomes much larger, and hence Δx increases. This will reduce Δp_x, thus making smaller values of p_x more probable and reducing the average value for p_x. A *decrease in kinetic energy* is the consequence and can thus be made responsible for the stabilising effect.

The virial theorem would – at a first glance – contradict this argument, as $\langle \hat{T} \rangle = -E$, and a smaller value of $\langle \hat{T}_{mol} \rangle$ than $\Sigma \langle \hat{T}_{atom} \rangle$ would thus result in a higher total energy – that is, destabilisation. Consequently, there must be another component in the binding process, and it is easily found when we re-optimise our function

$$|\psi_+\rangle = N(|a\rangle + |b\rangle) = \frac{1}{\sqrt{2(1+S)}}(e^{-\eta r_a} + e^{-\eta r_b})$$

with respect to η. Now, we will not find the energy minimum for $\eta = 1$ as for the hydrogen atom, but for $\eta = 1.25$, which means a *contraction* of the original functions. This contraction is an *intra-atomic* effect, occurring in the directions perpendicular to the bond axis, in y and z directions. Therefore, the amounts of Δy and Δz decrease, allowing higher p_y and p_z values and thus a compensation of the lowered kinetic energy in the bond direction, which is an *inter-atomic* effect. This contraction does not have much influence on the total energy, as can be seen from the blue curve in Figure 4.3, but it slightly shortens the bond distance. The related change in electron density distribution is visible in the blue curve of Figure 4.4.

We can thus summarise that the superposition of probability functions and the formation of a new, molecular probability function with different characteristics are the main foundations for a description of chemical bonding, and that lowering of the kinetic energy in the bond direction is the essential first step in the formation of a bond. This result shows the MO model to be much more appropriate for the description of molecules than the VB model, where the original atomic functions would retain their characteristics upon formation of a molecule or have to be forced to 'hybridise' in order to overcome the associated shortcomings.

As yet, we have dealt only with a one-electron system, and consequently all aspects of the chemical bond included so far are *one-electron contributions to the chemical bond*. When we move to larger molecules with more electrons, we will also have to consider the mutual repulsion of the electrons, which adds considerably to the complexity of the chemical bond.

4.5.2
Chemical Bonds in n-Electron Systems

The operator for the interaction of two electrons is the Coulomb operator $\frac{1}{r}$. We will first consider the interaction of two electrons in the charge distributions $\psi_i{}^2$ and $\psi_j{}^2$, which results in the integral $\int \psi_i{}^2(1)\frac{1}{r}\psi_j{}^2(2)\,d\tau_1\,d\tau_2$, which is usually written in the short notation $\langle i^2(1)/j^2(2)\rangle$. If we deal with two electrons distributed over two different functions each, $\psi_i{}^*\psi_j$ and $\psi_k{}^*\psi_l$, we must consider the four-center integrals for their repulsion

$$\int \psi_i{}^*(1)\psi_j(1)\frac{1}{r_{12}}\psi_k^*(2)\psi_l(2)\,d\tau_1\,d\tau_2 \equiv \langle ij/kl\rangle.$$

This is the so-called *Coulomb-Integral*. However, due to the antisymmetry principle, this electron repulsion is accompanied by another term, which is a consequence of the antisymmetry required for functions describing fermions; this second term is called the *Exchange Integral*, as it results from an exchange of two assigned functions as

$$\int \psi_i{}^*(1)\psi_k(1)\frac{1}{r_{12}}\psi_j^*(2)\psi_l(2)\,d\tau_1\,d\tau_2 \equiv \langle ik/jl\rangle.$$

The exchange term assures that two electrons of the same spin would not come too close to each other, thus reflecting the Pauli exclusion principle.

In Chapter 5, we will deal in detail with the construction of antisymmetrised n-electron functions from one-electron functions, and re-encounter these integral expressions. However, for the discussion of the contributions to the chemical bond, it is sufficient to know that we have to consider both Coulomb and exchange contributions within the framework of electron–electron repulsion. This repulsion usually reduces the stabilisation of a bond, and it can influence whether a compound's ground state is a high-spin or a low-spin electronic configuration – i.e., with unpaired or paired spins.

The approach to construct n-electron functions from a number of one-electron functions implies that the electrons can move independently of each other (with the restrictions provided for electrons with same spin by the exchange term). This simplification must be corrected in some cases, in particular for electrons with different spin, and the resulting term of correction is often considered as another contribution to the chemical bond in n-electron systems under the name of *electron correlation*, which again consists of integrals of the type $\langle ik/jl \rangle$. In fact, this contribution corresponds to a gain in energy, when the electrons 'avoid each other' as much as possible. Details and relevance of electron correlation will be discussed in connection with quantitative methods.

Another contribution to chemical bonding arises from the presence of different nuclei in the system. Different nuclei have different tendencies to release or to bind electrons, measurable through ionisation energies I_j and electron affinities E_j. Both quantities have been utilised by *Mulliken* to define the *electronegativity* of the elements as $EN = \frac{(I_j+E_j)}{2}$. The presence of elements with different electronegativity induces a *charge transfer* from and to these nuclei, thereby leading to uneven charge distributions, which cause local dipoles and multipoles; they are also responsible for the global dipole moment of a molecule. The interaction between the local partial charges is another contribution to the stability of chemical bonds (the extreme case would be the type of bonding classified as 'purely electrostatic') and can be quantified, allocating in a simplified procedure the varying charge distribution to fractional charges of the atoms in the molecule, by the simple electrostatic expression

$$\sum_i \sum_{<j} \frac{\delta q_i . \delta q_j}{r_{ij}}$$

where i and j refer to atomic centres.

These contributions to the chemical bond (and to the reactivity of a chemical compound!) are of particular importance, when one considers their long-range action, depending on $\frac{1}{r}$, compared to the exponentially decreasing overlap terms $\psi_i{}^* \psi_j$ in the integrals $\langle ik/jl \rangle$. At larger distances, such electrostatic forces will be dominating, and many methods in quantum chemistry make use of this fact in order to simplify the computational effort.

Finally, one must not forget the *relativistic contributions*, which play a considerable role in chemical bonding in heavier ($Z > 30$) elements. In a quantitative treatment of molecules these contributions are mostly included via corrections for the functions describing inner electrons of the heavier atoms. Although the main effects concern the inner electrons, as a consequence of them the valence shells are also influenced, and this leads to bond shortening and changes in the energies of electronic levels. The most 'shining' example is the unique yellow colour of metallic gold, which results from a relativistic contraction of the electron distribution and a subsequent reduction of the gap between electron energies in the valence region. Without this effect, gold would not be much different from silver in appearence.

4.5.3
Qualitative MO Models for Molecules

Having seen the manifold contributions in n-electron systems with different types of nuclei, it seems necessary to examine some more of the qualitative (even sometimes semiquantitative) models used in chemistry. These models are almost all based on one-electron theory, which means that they would neglect all effects arising from the presence of other electrons. Only a quantitative evaluation of these effects can tell us, how serious an error source this neglect is in every specific case, and to what extent the combination of the one-electron approximation with further approximations could eventually lead to a fortunate error compensation in a series of similar compounds.

A few indications towards the importance of electron–electron interaction in n-electron systems can help to demonstrate, however, that most of these models are highly questionable and misleading.

The example of high-spin and low-spin complexes has already been mentioned. Experimental measurements can easily detect the difference from the magnetic properties (dia- or paramagnetism), which indicate whether one or more electrons are present in an unpaired spin state. Simple models assuming d-orbitals for the central metal and their filling by two (paired) or single (unpaired) electrons are employed to interpret the measurements, and they even introduce a 'pairing energy' for the electrons 'residing' in the same spatial orbital with α or β spin as a determining factor, whether electrons should be 'paired' or a state with some 'unpaired' electrons is energetically favoured. The predicted paramagnetism usually does not coincide with the experimental data, and another artificial term has been introduced to 'explain' this, named 'orbital quenching'. We know, however, that neither l and s nor L and S are good quantum numbers in these complexes, and that consequently no 'd-orbitals' are available. Only the evaluation of a completely new probability function for the whole system in the form of a quantitative MO treatment can provide significant data for electronic and magnetic properties, and the difference in total energy of possible overall spin states decides, which state is the most stable. There is hardly any chance to model all the contributing effects in a qualitative way.

Another bad example of a model based on one-electron theory is the HOMO-LUMO concept, where the proceeding of reactions is 'explained' on the basis of an energy difference between the 'highest occupied' and 'lowest unoccupied' molecular orbital. Within the framework of a one-electron formalism these energy levels indeed remain unchanged, but as soon as we include electron–electron interaction, they can strongly change upon occupation or disoccupation, even to the effect of a new ranking of the orbitals on the energy scale. Therefore, models based on the one-electron approximations do neglect a large part of physics, and their putative success is mostly based on a lucky compensation of errors.

After having dealt systematically with all approximations in quantitative MO theory in Chapter 5, we will return to the question of useful qualitative concepts in chemistry again and show, what remains as a helpful and physically relevant approach to describe the binding and reactivity of molecules.

Test Questions Related to this Chapter

1. Is it correct to assign p electrons to the nitrogen in the pyridine molecule?
2. Which model has developed the hybridisation concept, and for what reason?
3. Which quantum numbers are 'good' quantum numbers for an atom/a molecule?
4. What were the reasons for introducing spin into quantum mechanics?
5. What are the differences between Fermions and Bosons?
6. For which configuration of a molecule is the virial theorem valid?
7. What are the one-electron, and the two-electron contributions to the chemical bond?
8. Why do we have electron correlation as a separate term in the two-electron contributions?
9. Which contributions to molecular stability and interaction are dominant at larger distances?

5
Approximations for Many-Electron Systems

> *Mathematics and natural science are specially distinguished from man's other intellectual activities by being teachable and verifiable.*
>
> Wolfgang Pauli, 20th century

In this chapter we will discuss the approximations needed to proceed from general vector space of matter to practicable methods for a quantitative computational treatment of chemical systems. For each of the approximations introduced, the associated possible error sources and possibilities to overcome potential errors will be briefly mentioned.

5.1
Non-Relativistic Stationary Systems

We start from an abstract probability vector in the vector space of matter, and proceed to its realisation in the form of a probability function for a system of nuclei and electrons, depending on the variables of space (R for the nuclei, r for the electrons) and time.

$$|A\rangle \quad \Rightarrow \quad |\Psi(R_i, r_i, t)^{rel}\rangle$$

The superscript 'rel' indicates that relativistic effects would have to be included. However, the first simplification we will make is to neglect such effects when we proceed. It has already been mentioned in the discussion of the chemical bond that this will cause problems when dealing with the chemistry of heavier elements. On the other hand, there are methods available to introduce the relativistic contributions at a later stage, and consequently we do not have to worry at this point.

The second simplification we accept is to deal with stationary systems, which means to eliminate time from the probability function. This implies that we can deal with dynamics only via a combination of solutions for the time-independent system with a time-evolution function or other formalisms introducing a dynam-

The Basics of Theoretical and Computational Chemistry. Edited by B. M. Rode, T. S. Hofer and M. D. Kugler

ISBN: 978-3-527-31773-8

ical transition between stationary states. By combining approximations we arrive at the Schrödinger equation of the following form:

$$\hat{H}|\Psi^{non-rel}(R_i, r_i)\rangle = E^{non-rel}|\Psi^{non-rel}(R_i, r_i)\rangle.$$

From this point, we will imply that $|\Psi\rangle$ is a non-relativistic function, without specifying it by the superscript 'non-rel'.

5.2
Adiabatic Approximation – The Born–Oppenheimer Approximation

The Schrödinger equation for nuclei and electrons, resulting from the previous two simplifications,

$$\hat{H}|\Psi(R_i, r_i)\rangle = E|\Psi(R_i, r_i)\rangle$$

still remains a highly complicated expression, and further simplifications are needed to reach the goal of a manageable task for a numerical solution. When considering the magnitude of the masses of nuclei and electrons, respectively, one could assume that the mobility of the electrons is so much faster than that of the nuclei that upon every move of the nuclei the electrons would immediately adapt, thus always being in an eigenstate of the system. This assumption, which allows a composition of the function $|\Psi(R_i, r_i)\rangle$ as a product of one for the nuclei and one for the electrons as

$$|\Psi(R_i, r_i)\rangle = |\Psi(R_i)\rangle \cdot |\Psi(r_i)\rangle$$

is called the *adiabatic approximation*. In a more colloquial way one can state that the electrons move independently of the nuclei due to the different masses, and in this form the simplification is known as the *Born–Oppenheimer (BO) approximation*. As a consequence, we now have two separate equations for the nuclei and for the electrons:

$$\hat{H}^{nucl}|\Psi(R_i)\rangle = E^{nucl}|\Psi(R_i)\rangle \qquad \hat{H}^{el}|\Psi(r_i)\rangle = E^{el}|\Psi(r_i)\rangle.$$

These approximations seem physically well-justified, but whenever two energy surfaces of different electronic states with different nuclear geometry come close to each other, one must be aware that coupling terms might become relevant, thus requiring a correction.

The separation of electronic and nuclear probability functions has a serious consequence for the theoretical treatment of molecular vibrations, reaction paths and geometry optimisations of molecules, i.e., with every aspect related to the displacement of nuclei on *energy surfaces*. The (multidimensional) dependence of total energy from a number of coordinates – this can be ordinary space coordinates or internal coordinates such as bond lengths and angles – is called energy (hyper)surface, and it is clear that after the separation of nuclear and electronic movements, such energy surfaces will have to be evaluated *pointwise* by solving the electronic Schrödinger equation for each nuclear configuration separately. The contribution of nuclear repulsion to the total energy for each of these configurations is added to the electronic energy by a simple electrostatic term summing over all nuclei N_I with the charges Z_I

$$E^{nuc} = \sum_I \sum_{<J} \frac{Z_I \cdot Z_J}{R_{IJ}}.$$

It is clear that this pointwise evaluation of energy surfaces is not a very elegant way to determine equilibrium geometries, vibrational frequencies and other data related to nuclear coordinates, but many of these procedures have been automated by algorithms, mostly gradient methods, implemented in quantum chemical programmes, and thus are less cumbersome than they appear at a first glance.

5.3 The Independent Particle Approximation

After all of the approximations introduced so far, the remaining task is to solve the electronic Schrödinger equation for the n-electron system

$$\hat{H}^{el}|\Psi(r_i)\rangle = E^{el}|\Psi(r_i)\rangle$$

and to evaluate the total energy by adding the nuclear repulsion term as $E^{total} = E^{el} + E^{nuc}$. However, a probability vector $|\Psi(r_i)\rangle$ for an n-electron system is a very complex item and one had to seek a method to construct it from simpler building blocks, as has already been done in the discussion of the chemical bond, where one separate function has been assigned to each electron. For an n-electron system, the total probability function would thus be formed by a product of n one-electron functions, called the *Hartree Product*:

$$|\Psi\rangle = \frac{1}{\sqrt{n!}}[\psi_1(1)\psi_2(2)\ldots\psi_n(n)].$$

Due to the physical meaning of such a product ansatz, implying that the concerned particles can move independently in the system, this approach is called the *Independent Particle Approximation (IPA)*. As it implies the assignment of one separate probability function or (molecular) ‘orbital’ to each electron, it is also referred to as the *Orbital Approximation*.

Although the space available for the electrons in a molecule appears very large compared to their size, the IPA is much more serious in terms of physics than the BO approximation, especially if electron density is 'concentrated' in 'narrower' regions of the molecule due to strong binding. In such cases, the electron movements are less independent of each other, but must proceed in a more correlated way. For this reason, the energy gain related to such a coordinated movement is called *'correlation energy'*, and so we now have a more precise definition of the *electron correlation* contribution to the chemical bond. At the same time, we understand that this is not a fundamentally new contribution to binding, but rather only a correction of a previously made error due to the IPA. Nevertheless, the literature sometimes treats correlation energy as if it were a specific part of the binding energy. An example is the formation of the F_2 molecule from two F atoms. In the latter, due to a weakly shielded nuclear charge the valence electron density is strongly contracted, which calls for a more correlated movement of the electrons. Upon formation of the molecule, much more space is available for the electrons, and this difference actually plays such an important role for the overall energy balance that the formation of this molecule is almost exclusively attributed to correlation energy. It does not appear correct, however, to classify the binding in F_2 as an effect of a special 'correlation contribution' to the chemical bond; one should rather state that the IPA is inadequate for the description of this molecule.

Later, we will see that calculations overcoming this error of the IPA by explicitly evaluating correlation energy contributions lead to a dramatic increase in computational effort. Fortunately, for the large majority of chemical compounds these contributions are very minor (<1% of total energy) and can thus be neglected in most cases.

5.4 Spin Orbitals and Slater Determinants

While forming the Hartree product in the IPA formalism, we have not yet considered another important requirement for the n-electron probability function: due to the Pauli antisymmetry principle, electrons are fermions, and only an antisymmetric function will be adequate, therefore, to describe a n-electron system. For this reason, one must find a way to build a function from the one-electron functions guaranteeing antisymmetric behaviour. Before doing so, let us define in some more detail the one-electron functions $|\psi_i\rangle$ (in our notation we will always distinguish n-electron functions and one-electron functions by denoting them by a capital Ψ and small ψ, respectively).

Each function $|\psi_i\rangle$ consists of a 'spatial' part, describing the radial and angular distribution, and a spin function $|\sigma\rangle$. This spin function can have two eigenvalues, $+\frac{1}{2}$ or $-\frac{1}{2}$. The associated spin function for the first case will be denoted as $|\alpha\rangle$, that for the second case as $|\beta\rangle$. A common notation is

$$\psi_i \equiv |\psi_i\rangle \cdot |\alpha\rangle \qquad \bar{\psi}_i \equiv |\psi_i\rangle \cdot |\beta\rangle.$$

Such orbital functions including the spin component are called *spin orbitals*, and they are the common building blocks for the total probability function $|\Psi\rangle$.

The functional form for $|\Psi\rangle$ ensuring antisymmetric behaviour is the determinantal form. This can be illustrated in a simple example with two electrons. The function

$$|\Psi\rangle = \frac{1}{\sqrt{2}}[\psi_1(1)\psi_2(2) - \psi_1(2)\psi_2(1)]$$

is antisymmetric, as can be shown from

$$\hat{P}_{12}|\Psi\rangle = \frac{1}{\sqrt{2}}[\psi_1(2)\psi_2(1) - \psi_1(1)\psi_2(2)] = -|\Psi\rangle$$

This function can be written more simply as determinantal function

$$|\Psi\rangle = \frac{1}{\sqrt{2}}\begin{vmatrix} \psi_1(1) & \psi_1(2) \\ \psi_2(1) & \psi_2(2) \end{vmatrix}$$

and analogously we can construct normalised determinantal functions for any n-electron system from n spin orbitals as

$$|\Psi\rangle = \frac{1}{\sqrt{n!}}\begin{vmatrix} \psi_1(1) & \psi_1(2) & \cdots & \cdots & \psi_1(n) \\ \psi_2(1) & \psi_2(2) & \cdots & \cdots & \psi_2(n) \\ \cdots & \cdots & \cdots & \cdots & \cdots \\ \cdots & \cdots & \cdots & \cdots & \cdots \\ \psi_n(1) & \psi_n(2) & \cdots & \cdots & \psi_n(n) \end{vmatrix}$$

These determinantal functions are called *Spin Determinants* or – after their 'inventor' – *Slater Determinants.* According to an excess of either α or β spin, a total spin $S \neq \emptyset$ can result for the spin determinant, which is usually given by indicating the multiplicity $M = 2S + 1$ as a superscript on the left of the function symbol as $|^M\Psi\rangle$. For the sake of a simpler notation one often writes down only the diagonal of the determinant enclosed in two bars as

$$|^M\Psi\rangle = \frac{1}{\sqrt{n!}}|\psi_1(1)\psi_2(2)\ldots\psi_n(n)|$$

and we will make immediate use of this notation to explain a special case, the so-called closed-shell systems, where each spatial orbital appears combined with both α and β spin ('paired electrons'). Consequently, the multiplicity of such a system, which is representative for the ground state of the large majority of molecules, is 1. Thus, we can write the corresponding determinantal function as

$$|^1\Psi\rangle^{closed-shell} = \frac{1}{\sqrt{n!}}|\psi_1(1)\bar{\psi}_1(2)\psi_2(3)\bar{\psi}_2(4)\ldots\psi_{\frac{n}{2}}(n-1)\bar{\psi}_{\frac{n}{2}}(n)|$$

The next step to solve the Schrödinger equation for the n-electron system is to construct Slater determinants in a way suitable for a numerical solution, which means finding appropriate expressions for the spin orbitals in a basis of simpler functions or, in other words, in a finite vector space spanned by basis vectors suitable for this purpose. This is done by using the so-called LCAO-MO approach.

5.5
Atomic and Molecular Orbitals: The LCAO-MO Approach

Even one-electron spin orbitals for a molecule are highly complex and must be composed in an optimised way from basis functions. In the quantitative molecular orbital approach this is achieved, (as outlined in Chapter 2) by a linear combination of basis functions $\mathbf{\Psi} = \mathbf{\Phi}\mathbb{C}$, where every individual spin orbital $|\psi_i\rangle$ is formed with the basis vectors $|\varphi_j\rangle$, using the vector $\mathbb{C}_j$, containing the linear combination coefficients c_{ij}. The conventional (and in most cases most reasonable) choice of basis functions is the assignment of *'atomic orbitals (AO)'* to each atom in the system, representing the types of electron found in the period, where this atom stands in the periodic system. Later, we will see that a good description of molecules requires a larger set of such atomic orbitals, also including 'higher' functions, for example, p-functions for hydrogen and d-functions for carbon, nitrogen, and oxygen. After one has assigned a well-balanced number of such AOs to each atom, they are combined via the coefficients to *'molecular orbitals (MO)'*, and thus the method is termed by the acronym *LCAO-MO*, meaning a *'Linear Combination of Atomic Orbitals to Molecular Orbitals'*. It is evident that one should use the 'best possible' coefficients within the given basis set for this construction, and we will soon show how this is achieved using the variational principle.

The choice of an appropriate *basis set* is a most crucial task for this method. On the one hand, one wishes to give the electrons a maximum of flexibility to distribute over the system, which implies the use of a large number of different basis functions; on the other hand, the computational effort increases exponentially with the number of functions. This balance between accuracy and feasibility has become much less stringent through the continuous development of high-performance computing facilities, but it has only shifted to larger chemical systems, where this 'Scylla and Charybdis' situation is still encountered.

The types of function used as AOs mostly belong to three classes. As the solution of the Schrödinger equation for hydrogen produces rather complicated radial functions, *Slater* proposed to replace the radial part by nodeless functions of similar appearance, multiplied with the angular function part we already know as 'spherical harmonics'

$$\varphi_{STO} = N \cdot r^{n-1} \cdot e^{-\zeta \cdot r} \cdot Y_{\theta,\varphi}^{l,m}$$

Such functions are called *'Slater-Type Orbitals (STO)'*, and the exponents ζ in the e-function were optimised by Slater for various values of n, the main quantum number, which appears in the radial part of the function.

Although Slater functions apparently allow good MO constructions with a rather small basis set, they lead to numerical problems during integration, especially for functions with larger n. Therefore, the majority of computational chemists have decided to use another basis function type, called *Gauss-Type Orbitals (GTO)* or Gaussian functions, having the following form

$$\varphi_{GTO} = N \cdot e^{-\alpha \cdot r^2} \cdot Y_{\theta,\varphi}^{l,m}$$
$$\varphi_{GTO}^{nl} = N \cdot r^{2n-l-2} \cdot e^{-\alpha \cdot r^2} \cdot Y_{\theta,\varphi}^{l,m}$$

Such functions are easy to handle in the integral part of the programmes but, due to their lesser similarity to the radial functions resulting for hydrogen, one needs about three GTOs to obtain a comparable quality of description as with one STO. This 3 for 1 rule has become a (low-side) standard in the construction of the various basis sets available for atoms throughout the periodic system.

A very simple way to construct basis sets avoiding the angular part of the functions is to use '*Gaussian Lobe Orbitals (GLO)*' of the form $\varphi_{GLO} = N \cdot e^{-\alpha \cdot r^2}$ and to compose p-like functions by placing two such functions with + and − signs above and below the xy-, xz-, and yz planes of the local atomic coordinate system. Similarly, one can construct d-orbitals from four lobe functions. In this approach, the total number of functions increases considerably, but they are numerically the most convenient ones. Nowadays, neither STO nor GLO basis sets are used to a large extent, and GTOs have become the most frequently used standard of quantum chemical calculations.

After a calculation has been performed with a given basis set, further calculations with more extended basis sets can serve as a method-inherent control of the quality of the results. Total energy, virial quotient $\frac{\langle \hat{V} \rangle}{\langle \hat{T} \rangle}$ and the reproduction of physico-chemical properties can serve as quality criteria.

5.6 Quantitative Molecular Orbital Calculations

We will now transform the concept of the LCAO-MO method into a quantitative formalism, serving the purpose of a direct translation into computer program language. Besides a preparation for a quantitative computational treatment of chemical systems within the framework of quantum theory, the development of this formalism will give us a deeper physical understanding of widely used terms and show us some limitations of the method.

5.6.1 Calculations with Slater Determinants

As it has become clear that we must describe the n-electron system by Slater determinants, we will summarise the basic rules for calculations with such determinants, as this will greatly facilitate the handling of the expressions in the following.

The Hamiltonian of the electronic Schrödinger equation consists of two parts, namely a one-electron part (called the core Hamiltonian) and a two-electron part. It will make all further expressions simpler, if we change to atomic units from this point, where the mass and charge of the electron and $\hbar$ are all units. Thus, the Hamiltonian can be written as

$$\hat{H}^{el} = \hat{H}^{c} + \hat{G}^{ij} = \sum_{i} \hat{h}_{i}^{c} + \sum_{i} \sum_{<j} \frac{1}{r_{ij}}$$

where – following the use of one-electron functions – $\hat{H}^{c}$ has been written as a sum of one-electron operators $\hat{h}_{i}^{c}$ defined as

$$\hat{h}_{i}^{c} = -\frac{1}{2} \Delta_{i} + \sum_{N} \frac{Z_{N}}{R_{iN}}$$

which each contain the kinetic energy of the ith electron and its interaction with all N nuclei of the system. The two-electron term of the Hamiltonian, summing up over all pairs of electrons (i, j), describes the interaction between the electrons.

We will thus have to deal with three types of scalar products/integrals while solving the eigenvalue equation of the Hamiltonian, namely the overlap integral $\langle\Psi|\Psi\rangle$, integrals containing a one-electron operator and those containing a two-electron operator.

Now we will summarise the results of these integrations:

5.6.1.1 Overlap Integrals

As our basic determinant we assume

$$|\Psi\rangle = |\psi_{1}(1)\psi_{2}(2) \ldots \psi_{n}(n)|$$

and the scalar product (overlap integral) of this function with itself $\langle\Psi|\Psi\rangle$ results as n!, from which we can deduce the normalisation factor for the function, $\frac{1}{\sqrt{n!}}$. If we have two determinants that differ in only one of the spin orbitals:

$$|\Psi_{k}\rangle = |\psi_{1}(1)\psi_{2}(2) \ldots \psi_{k}(k) \ldots \psi_{n}(n)|$$
$$|\Psi_{l}\rangle = |\psi_{1}(1)\psi_{2}(2) \ldots \psi_{l}(k) \ldots \psi_{n}(n)|$$

the scalar product (overlap integral) vanishes, i.e., $\langle\Psi_{k}|\Psi_{l}\rangle = \emptyset$. The same is of course true for spin determinants that differ in two or more spin orbitals.

5.6.1.2 Integrals of One-Electron Operators

For two identical spin determinants, the scalar product (integral) results as

$$\langle\Psi|\hat{h}^{c}\Psi\rangle = \sum_{i} \langle\psi_{i}|\hat{h}^{c}\psi_{i}\rangle' \equiv \sum_{i} h_{ii}{}'$$

The latter is just a short notation for such integrals, which will prove convenient for the further formalism. We must sum these elements for all spin orbitals in the determinant, and the 'prime' sign indicates that we still have to consider the spin state of the determinant by performing spin integration.

If the two spin determinants differ in one spin orbital, we obtain

$$\langle \Psi_k | \hat{h}^c \Psi_l \rangle = \langle \psi_k | \hat{h}^c \psi_l \rangle' \equiv h'_{kl}$$

Only a single element containing the two different spin orbitals remains. If the two determinants differ in two or more spin orbitals, the scalar product (integral) vanishes:

$$\langle \Psi_{kl} | \hat{h}^c \Psi_{mn} \rangle = \emptyset$$

5.6.1.3 Integrals of Two-Electron Operators

In this part, we will use the previously introduced notation

$$\int \psi_i^*(1)\psi_j(1)\frac{1}{r_{12}}\psi_k^*(2)\psi_l(2)\,d\tau_1\,d\tau_2 \equiv \langle ij/kl \rangle$$

for two identical spin determinants, the scalar product (integral) results as

$$\left\langle \Psi \middle| \frac{1}{r_{12}} \Psi \right\rangle = \sum_i \sum_{<j} (J'_{ij} - K'_{ij})$$

with $J_{ij} \equiv \langle ij/ij \rangle$ and $K_{ij} \equiv \langle ij/ji \rangle$ as short notations for a Coulomb and an exchange integral.

If the two spin determinants differ in one spin orbital, we obtain

$$\left\langle \Psi_k \middle| \frac{1}{r_{12}} \Psi_l \right\rangle = \sum_{j \neq k} [\langle kj/lj \rangle' - \langle kj/jl \rangle']$$

If the determinants differ in two spin orbitals, the result of the scalar product (integral) is

$$\left\langle \Psi_{kl} \middle| \frac{1}{r_{12}} \Psi_{mn} \right\rangle = [\langle km/ln \rangle' - \langle km/nl \rangle'] \quad \forall \sigma_k = \sigma_m \; and \; \sigma_l = \sigma_n$$

Finally, if they differ in more than two spin orbitals, the scalar product (integral) becomes zero

$$\left\langle \Psi_{k!m} \middle| \frac{1}{r_{12}} \Psi_{nop} \right\rangle = \emptyset$$

Again, in all of the above-listed expressions, the 'prime' sign indicates that we still have to perform spin integration.

With these rules for calculations with spin determinants we are now well prepared to find a formulation of the Schrödinger equation

$$\mathbb{H}^{\varphi}\mathbb{C} = \mathbb{S}\mathbb{C}\mathbb{E}$$

obtained earlier in Chapter 2, within the framework of the LCAO-MO formalism using spin orbitals combined to a Slater determinant.

5.6.2
The Hartree–Fock Method

Let us return to the Hamiltonian

$$\hat{H} = \hat{H}^c + \hat{G}^{ij} = \sum_i \hat{h}_i^c + \sum_i \sum_{<j} \frac{1}{r_{ij}}$$

and calculate its expectation value with a closed-shell spin determinant

$$|\Psi\rangle = \frac{1}{\sqrt{n!}} |\psi_1(1)\bar{\psi}_1(2)\psi_2(3)\bar{\psi}_2(4)\ldots\psi_{\frac{n}{2}}(n-1)\bar{\psi}_{\frac{n}{2}}(n)|$$

Applying the rules for calculation with determinantal functions we obtain

$$\langle\hat{H}\rangle = \langle\Psi|\hat{H}\Psi\rangle = \sum_i^{n/2} h_{ii}{}' + \sum_i^{n/2} \sum_{<j}^{n/2} (J'_{ij} - K'_{ij})$$

where the summations run over n/2 spatial orbitals. The fact that all of them occur twice, with α and β spin, will be taken into account by the spin integration to be performed now. The result for $h_{ii}{}'$ is obviously 2 h_{ii}, as both spin integrations $\langle\alpha|\alpha\rangle$ and $\langle\beta|\beta\rangle$ give the value 1. For the elements $J_{ij}{}'$ and $K_{ij}{}'$ we will have to take a closer look at the possible spin combinations for the Coulomb integrals

$$\int \psi_i{}^*(1)\psi_j{}^*(2)\frac{1}{r_{12}}\psi_i(1)\psi_j(2)\, d\tau_1\, d\tau_2$$

and the exchange integrals

$$\int \psi_i{}^*(1)\psi_j{}^*(2)\frac{1}{r_{12}}\psi_j(2)\psi_i(1)\, d\tau_1\, d\tau_2$$

$$i\ \alpha\ j\ \alpha\ /\ i\ \alpha\ j\ \alpha = J_{ij}$$

$$i\ \alpha\ j\ \beta\ /\ i\ \alpha\ j\ \beta = J_{ij}$$

$$i\ \beta\ j\ \alpha\ /\ i\ \beta\ j\ \alpha = J_{ij}$$

$$i\ \beta\ j\ \beta\ /\ i\ \beta\ j\ \beta = J_{ij}$$

represents the four possible combinations for the elements of the Coulomb operator. For the exchange operator these combinations are:

$$i\ \alpha\ j\ \alpha\ /\ j\ \alpha\ i\ \alpha = K_{ij}$$
$$i\ \alpha\ j\ \beta\ /\ j\ \alpha\ i\ \beta = \emptyset$$
$$i\ \beta\ j\ \alpha\ /\ j\ \beta\ i\ \alpha = \emptyset$$
$$i\ \beta\ j\ \beta\ /\ j\ \beta\ i\ \beta = K_{ij}.$$

All combinations of spatial and spin functions lead to a value J_{ij}, but only for the all-α and all-β cases do we obtain K_{ij}, as in the other two cases the spin integration $\langle\alpha|\beta\rangle = \emptyset$ annihilates the terms. Thus, the result of spin integration for the two-electron terms gives $4J_{ij} + 2K_{ij}$.

In the closed-shell case we still have to count the possible terms $J_{ii} \equiv K_{ii}$. They are only possible if ψ_i is combined once with $|\alpha\rangle$ and once with $|\beta\rangle$, and thus $J_{ii}{}' = 2J_{ii}$, or – useful for the inclusion in the following equation – $J_{ii}{}' = J_{ii} + K_{ii}$. This will make the summation easier, as we will count over both indices i and j, allowing also i = j, and arrive at the final expression for the expectation value of a closed-shell determinant, expressed in spatial orbitals:

$$\langle\hat{H}\rangle = 2\sum_{i}^{n/2} h_{ii} + \sum_{i}^{n/2}\sum_{j}^{n/2}(2J_{ij} - K_{ij}) = E_{el}$$

which delivers the value for the electronic energy. To obtain the total energy, one has to add the nuclear repulsion energy

$$E_{nuc} = \sum_{I}\sum_{<J}\frac{Z_I \cdot Z_J}{R_{IJ}}.$$

It is also common to define one-electron energies ε_i and to use them to obtain a short expression for the electronic energy as

$$\varepsilon_i = h_{ii} + \sum_{j}^{n/2}(2J_{ij} - K_{ij}) \qquad E_{el} = \sum_{i}(\varepsilon_i + h_{ii})$$

The values ε_i are one-particle energies, also called *'orbital energies'*, and correspond to a one-particle eigenvalue equation

$$\left[\hat{h}_i + \sum_{j}(2\hat{J}_j - \hat{K}_j)\right]|\psi_i\rangle = \varepsilon_i|\psi_i\rangle \qquad \text{or} \qquad \hat{F}_i|\psi_i\rangle = \varepsilon_i|\psi_i\rangle$$

after the introduction of the so-called Fock operator $\hat{F}$

$$\hat{F} = \hat{h}_c + \sum_j (2\hat{J}_j - \hat{K}_j)$$

consisting of the core Hamiltonian plus a sum of Coulomb and exchange operators, the detailed forms of which we will encounter shortly.

At this point a strong warning must be issued: the 'orbital energies' ε_i are by no means physical observables, rather they are an artifact produced by our independent particle approximation, and if they are often utilised for arguments or 'explanations' in chemistry, even on the basis of quantitative MO calculations, they do not become more 'real' ... they are pseudo-eigenvalues and – in the best case – a model to imagine the various energy states of electrons in an n-electron system in order to describe some aspects of electronic spectra.

Let us return to the previously derived energy expression for a closed-shell system

$$E_{el} = 2\sum_i^{n/2} h_{ii} + \sum_i^{n/2}\sum_j^{n/2} (2J_{ij} - K_{ij})$$

which contains all spatial orbitals $|\psi_i\rangle$. As mentioned above, these complex functions will be constructed from a basis set of atomic orbitals $\{\varphi_j\}$ with the linear combination coefficients c_{ij}. The next task is to find the best possible form of the spin orbitals. Here, the LCAO approach provides a good method to optimise all spin orbitals via the coefficients and the *variational principle*, optimising the electronic energy, for which at the minimum the condition $\delta E = \emptyset$ should be fulfilled – that is,

$$\delta E = 2\sum_i^{n/2} \delta h_{ii} + \sum_i^{n/2}\sum_j^{n/2} (2\delta J_{ij} - \delta K_{ij}) = \emptyset$$

This method – to solve the electronic Schrödinger equation based on an antisymmetrised Hartree product of one-electron functions (i.e., for a Slater determinant as total function $|\Psi\rangle$ by means of the variational principle) – is called the *Hartree–Fock method*, and we will now show, how the variation is performed in the framework of the LCAO-MO approach.

5.6.3 Hartree–Fock Calculations in the LCAO-MO Approach: The Roothaan–Hall Equation

When all spin orbitals $|\psi_i\rangle$ are expressed as a linear combination of the basis functions $|\varphi_j\rangle$, we can express this in matrix form as $\boldsymbol{\Psi} = \boldsymbol{\Phi} \cdot \mathbb{C}$, and the integral involving two spin orbitals results as

$$\langle\psi_i|\psi_j\rangle = \mathbb{C}_i^+ \langle\boldsymbol{\Phi}|\boldsymbol{\Phi}\rangle \mathbb{C}_j = \mathbb{C}_i^+ \mathbb{S} \mathbb{C}_j$$

For scalar products, where an operator is involved, we obtain

$$\langle \psi_i | \hat{O} \psi_j \rangle = \mathbb{C}_i^+ \langle \mathbf{\Phi} | \hat{O} \mathbf{\Phi} \rangle \mathbb{C}_j = \mathbb{C}_i^+ \mathbb{O}^{\varphi} \mathbb{C}_j$$

We will apply this to our expression for the closed-shell energy

$$E_{el} = 2 \sum_i^{n/2} h_{ii} + \sum_i^{n/2} \sum_j^{n/2} (2J_{ij} - K_{ij})$$

and for this purpose we have to define the Coulomb and the exchange operator explicitly, by showing their action on a spin orbital:

$$\hat{J}_i | \psi_k(\mu) \rangle = \int \frac{\psi_i^*(\nu) \psi_i(\nu)}{r_{\mu\nu}} d\tau_\nu | \psi_k(\mu) \rangle$$

$$\hat{K}_i | \psi_k(\mu) \rangle = \int \frac{\psi_i^*(\nu) \psi_k(\nu)}{r_{\mu\nu}} d\tau_\nu | \psi_i(\mu) \rangle$$

where μ and ν denote the two electrons involved. From these expressions for the two integral operators we recognise that we should actually know the spin orbitals – and thus the solutions for the optimal coefficients! – in order to formulate the operators. As everybody will guess, this will in practice lead to an iterative procedure.

The closed-shell energy can now be written as

$$E_{el} = 2 \sum_i \mathbb{C}_i^+ \mathbb{H}^c \mathbb{C}_i + \sum_i \sum_j (2\mathbb{C}_j^+ \mathbb{J}_i \mathbb{C}_j - \mathbb{C}_j^+ \mathbb{K}_i \mathbb{C}_j)$$

Due to the double summation over i and j in the second term, we can exchange the indices and thus arrive at a simpler expression

$$E_{el} = \sum_i \mathbb{C}_i^+ \left(2\mathbb{H}^c + 2 \sum_j \mathbb{J}_j + \sum_j \mathbb{K}_j \right) \mathbb{C}_i$$

Those readers who are not interested in the detailed derivation of the variational equations may jump from this point to the Roothaan–Hall equation, which is the final result. However, for those who are interested in finding how to reach this equation, we show it in detail here.

In order to adapt the variational expression

$$\delta E = 2 \sum_i \delta h_{ii} + \sum_i \sum_j (2\delta J_{ij} - \delta K_{ij}) = \emptyset$$

to the LCAO-MO formalism, we must first determine the terms δh_{ii}, δJ_{ij} and δK_{ij} in the basis $\{\varphi_i\}$:

$$\delta h_{ii} = \delta \mathbb{C}_i^+ \mathbb{H}^c \mathbb{C}_i + \mathbb{C}_i^+ \mathbb{H}^c \delta \mathbb{C}_i = \delta \mathbb{C}_i^+ \mathbb{H}^c \mathbb{C}_i + \delta \mathbb{C}_i^T \mathbb{H}^{c*} \mathbb{C}_i^*$$

The reformulation of the second term in this expression facilitates the final collection of terms and will also be executed for all other terms.

The terms δJ_{ij} and δK_{ij} are more complex than the term δh_{ii}, as the operators contain (as we have shown previously) the coefficients of the spin orbitals and, therefore, must be varied themselves:

$$\begin{aligned}\delta J_{ij} &= \delta \mathbb{C}_i^+ \mathbb{J}_j \mathbb{C}_i + \mathbb{C}_i^+ \mathbb{J}_j \delta \mathbb{C}_i + \mathbb{C}_i^+ \delta \mathbb{J}_j \mathbb{C}_i \\ &= \delta \mathbb{C}_i^+ \mathbb{J}_j \mathbb{C}_i + \delta \mathbb{C}_i^T \mathbb{J}_j^* \mathbb{C}_i^* + \mathbb{C}_i^+ \delta \mathbb{J}_j \mathbb{C}_i\end{aligned}$$

The term $\mathbb{C}_i^+ \delta \mathbb{J}_j \mathbb{C}_i$ has now to be resolved, resulting in

$$\mathbb{C}_i^+ \delta \mathbb{J}_j \mathbb{C}_i = \delta \mathbb{C}_j^+ \mathbb{J}_i \mathbb{C}_j + \mathbb{C}_j^+ \mathbb{J}_i \delta \mathbb{C}_j = \delta \mathbb{C}_j^+ \mathbb{J}_i \mathbb{C}_j + \delta \mathbb{C}_j^T \mathbb{J}_i^* \mathbb{C}_j^*$$

This leads us to the final expression

$$\delta J_{ij} = \delta \mathbb{C}_i^+ \mathbb{J}_j \mathbb{C}_i + \delta \mathbb{C}_i^T \mathbb{J}_j^* \mathbb{C}_i^* + \delta \mathbb{C}_j^+ \mathbb{J}_i \mathbb{C}_j + \delta \mathbb{C}_j^T \mathbb{J}_i^* \mathbb{C}_j^*$$

and we obtain the analogous expression for the variation of the K_{ij} elements

$$\delta K_{ij} = \delta \mathbb{C}_i^+ \mathbb{K}_j \mathbb{C}_i + \delta \mathbb{C}_i^T \mathbb{K}_j^* \mathbb{C}_i^* + \delta \mathbb{C}_j^+ \mathbb{K}_i \mathbb{C}_j + \delta \mathbb{C}_j^T \mathbb{K}_i^* \mathbb{C}_j^*.$$

The terms for all variations δh_{ii}, δJ_{ij} and δK_{ij} will be used for the expression of δE

$$\begin{aligned}\delta E = 2 \sum_i \delta \mathbb{C}_i^+ \mathbb{H}^c \mathbb{C}_i &+ \sum_i \sum_j 2 \delta \mathbb{C}_i^+ \mathbb{J}_j \mathbb{C}_i - \sum_i \sum_j \delta \mathbb{C}_i^+ \mathbb{K}_j \mathbb{C}_i \\ &+ \sum_j \sum_i 2 \delta \mathbb{C}_j^+ \mathbb{J}_i \mathbb{C}_j - \sum_j \sum_i \delta \mathbb{C}_j^+ \mathbb{K}_i \mathbb{C}_j + 2 \sum_i \delta \mathbb{C}_i^T \mathbb{H}^{c*} \mathbb{C}_i^* \\ &+ \sum_i \sum_j 2 \delta \mathbb{C}_i^T \mathbb{J}_j^* \mathbb{C}_i^* - \sum_i \sum_j \delta \mathbb{C}_i^T \mathbb{K}_j^* \mathbb{C}_i^* \\ &+ \sum_j \sum_i 2 \delta \mathbb{C}_j^T \mathbb{J}_i^* \mathbb{C}_j^* - \sum_j \sum_i \delta \mathbb{C}_j^T \mathbb{K}_i^* \mathbb{C}_j^*\end{aligned}$$

We can now exchange the indices in the summations $\sum_j \sum_i$ and the corresponding elements and thus collect all terms with $\delta \mathbb{C}_i^+$ and $\delta \mathbb{C}_i^T$, which gives

$$\begin{aligned}\delta E = 2 \sum_i \delta \mathbb{C}_i^+ &\left[\mathbb{H}^c + \sum_j (2\mathbb{J}_j - \mathbb{K}_j) \right] \mathbb{C}_i \\ + 2 \sum_i \delta \mathbb{C}_i^T &\left[\mathbb{H}^{c*} + \sum_j (2\mathbb{J}_j^* - \mathbb{K}_j^*) \right] \mathbb{C}_i^*\end{aligned}$$

or with the use of the previously defined Fock operator

$$\hat{F} = \hat{h}_c + \sum_j (2\hat{J}_j - \hat{K}_j)$$

$$\delta E = 2\sum_i \delta \mathbb{C}_i^+ \mathbb{F} \mathbb{C}_i + 2\sum_i \delta \mathbb{C}_i^T \mathbb{F}^* \mathbb{C}_i^*$$

During the performance of the variation one needs to maintain the orthonormality of the spin orbitals $|\psi_i\rangle$. In terms of the LCAO-MO formalism this means

$$\delta(\mathbb{C}_i^+ \mathbb{S} \mathbb{C}_j) = \delta \mathbb{C}_i^+ \mathbb{S} \mathbb{C}_j + \mathbb{C}_i^+ \mathbb{S} \delta \mathbb{C}_j = \delta \mathbb{C}_i^+ \mathbb{S} \mathbb{C}_j + \delta \mathbb{C}_j^T \mathbb{S}^* \mathbb{C}_i^* = \emptyset$$

We have a total of $\frac{n}{2}$ (spatial) spin orbitals and thus $(\frac{n}{2})^2$ orthonormality conditions

$$2\sum_i \sum_j \delta \mathbb{C}_i^+ \mathbb{S} \mathbb{C}_j + 2\sum_i \sum_j \delta \mathbb{C}_i^T \mathbb{S}^* \mathbb{C}_j^*$$

which we can add with the method of Lagrange multipliers ε_{ij} to our equation for δE:

$$\delta E = 2\sum_i \delta \mathbb{C}_i^+ \left(\mathbb{F} \mathbb{C}_i - \sum_j \mathbb{S} \mathbb{C}_j \varepsilon_{ij} \right)$$
$$+ 2\sum_i \delta \mathbb{C}_i^T \left(\mathbb{F}^* \mathbb{C}_i^* - \sum_j \mathbb{S}^* \mathbb{C}_j^* \varepsilon_{ij} \right) = \emptyset$$

As this condition must hold for all variations $\delta \mathbb{C}_i^+$ and $\delta \mathbb{C}_i^T$, the expressions in the brackets must both be zero and thus lead to the equations

$$\mathbb{F} \mathbb{C}_i = \sum_j \mathbb{S} \mathbb{C}_j \varepsilon_{ij} \qquad \mathbb{F}^* \mathbb{C}_i^* = \sum_j \mathbb{S}^* \mathbb{C}_j^* \varepsilon_{ij}$$

If we subtract the complex conjugate of the second equation from the first, we obtain $\sum_j \mathbb{S} \mathbb{C}_j (\varepsilon_{ij} - \varepsilon_{ij}^*) = \emptyset$, from which follows that $\varepsilon_{ij} = \varepsilon_{ij}^*$, i.e., the matrix $\mathbb{E}$ of all ε_{ij} is a unitary matrix. Thus, we obtain a single equation for all i, j:

$$\mathbb{F} \mathbb{C} = \mathbb{S} \mathbb{C} \mathbb{E}$$

This is the famous *Roothaan–Hall equation*, the basis for the so-called *ab initio MO calculations.*

In order to solve the equation, one must perform a Löwdin orthonormalisation procedure first to convert the $\mathbb{S}$-matrix to the unit matrix, then to build the $\mathbb{F}$-matrix and to diagonalise it to obtain the eigenvalue matrix $\mathbb{E}$ and the eigenvector matrix $\mathbb{C}$.

We will now take a closer look at the elements of the $\mathbb{F}$-matrix, as this is helpful to understand the underlying physics and the steps of the calculation procedure.

A single element F_{rs} of the matrix $\mathbb{F}$, which is the representation of the Fock operator in the basis of atomic orbitals $\{\varphi_i\}$ is given by

$$F_{rs} = \langle\varphi_r(\mu)|\hat{F}\varphi_s(\nu)\rangle = \langle\varphi_r(\mu)|\hat{h}^c\varphi_s(\nu)\rangle + 2\sum_j J_{jrs} - \sum_j K_{jrs}$$

where μ and ν denote the two electrons involved, and J_{jrs} and K_{jrs} are abbreviations for the elements of the Coulomb and the exchange operator

$$J_{jrs} = \langle\varphi_r(\mu)|\hat{J}_j\varphi_s(\nu)\rangle \quad K_{jrs} = \langle\varphi_r(\mu)|\hat{K}_j\varphi_s(\nu)\rangle$$

We will now expand the element J_{jrs} by introducing $|\psi_j\rangle = \mathbf{\Phi}\mathbb{C}_j$ and the definition of the operator $\hat{J}_j$

$$\hat{J}_i|\psi_k(\mu)\rangle = \int \frac{\psi_i^*(\nu)\psi_i(\nu)}{r_{\mu\nu}} d\tau_\nu |\psi_k(\mu)\rangle$$

The element J_{jrs} then becomes

$$J_{jrs} = \int \varphi_r^*(\mu)\mathbb{C}_j^+ \frac{\langle\mathbf{\Phi}(\nu)|\mathbf{\Phi}(\nu)\rangle}{r_{\mu\nu}} \mathbb{C}_j\varphi_s(\mu)\, d\tau_\mu$$

or, after extracting the coefficient matrices and explicitly defining the integration of the overlap matrix

$$J_{jrs} = \mathbb{C}_j^+ \iint \varphi_r^*(\mu)\varphi_s(\mu) \frac{\mathbf{\Phi}^+(\nu)\mathbf{\Phi}(\nu)}{r_{\mu\nu}} d\tau_\mu\, d\tau_\nu \mathbb{C}_j$$

We can now replace the matrix multiplications by the corresponding summation terms

$$\mathbf{\Phi}^+(\nu)\mathbf{\Phi}(\nu) = \sum_t \sum_u \varphi_t^*(\nu)\varphi_u(\nu); \quad \mathbb{C}_j^+\mathbb{C}_j = \sum_t \sum_u c_{tj}^* c_{uj}$$

and with the notation

$$\iint \varphi_r{}^*(\mu)\varphi_s(\mu)\frac{1}{r_{\mu\nu}}\varphi_t^*(\nu)\varphi_u(\nu)\, d\tau_\mu\, d\tau_\nu \equiv (rs/tu)$$

we obtain the final expression for J_{jrs}, and by a completely analogous procedure we can also derive the expression for K_{jrs}

$$J_{jrs} = \sum_t \sum_u c_{tj}^* c_{uj}(rs/tu) \quad K_{jrs} = \sum_t \sum_u c_{tj}^* c_{uj}(rt/su)$$

Thus, one writes the complete expression for a Fock matrix element, performing summation over all occupied spin orbitals j as

$$F_{rs} = H_{rs}^c + \sum_j^{occ} \sum_t \sum_u c_{tj}^* c_{uj}[2(rs/tu) - (rt/su)]$$

The elements $\sum_j^{occ} c_{tj}^* c_{uj}$ are commonly defined as P_{tu}, and the whole set of them forms the so-called density matrix $\mathbb{P}$. With the density matrix elements we can re-write the Fock matrix elements in their final form

$$F_{rs} = H_{rs}^c + \sum_t \sum_u P_{tu}[2(rs/tu) - (rt/su)]$$

In the specific closed-shell case, where all spatial orbitals occur twice, the spatial orbitals can be regarded as doubly occupied, and the density matrix elements can be defined as $\sum_j^{n/2} 2c_{tj}^* c_{uj}$, which gives for the F_{rs} elements the form

$$F_{rs} = H_{rs}^c + \sum_t \sum_u P_{tu}\left[(rs/tu) - \frac{1}{2}(rt/su)\right]$$

This form now allows us to discuss the physical meaning of the constituents of the equation and to describe the practical process of construction of the Fock matrix representation in the basis set chosen.

The first term, H_{rs}^c, accounts for the kinetic energy of the electron and its interaction with all nuclei of the system. In the brackets of the second term, all Coulomb and exchange integrals are contained in the form of 1-, 2-, 3- and 4-center expressions [i.e. (rr/rr), (rr/tt), (rt/rt), (rr/tu), and (rs/tu)]. These contributions of electron–electron interactions are modulated, however, by the actual composition of the spin orbitals via the linear combination coefficients contained in the density matrix, and by the occupation of the spin orbitals. Thus, one can interpret the second term of the elements as the interaction of one electron with a 'field' of the $(n-1)$ other electrons.

At the same time it is evident that one needs to know the coefficients (i.e., the eigenvectors) in order to construct $\mathbb{P}$ and with it the F_{rs} elements. In other words, one needs the solution before starting the calculations. It is clear that this dilemma can only be solved in an iterative procedure, where a first set of coefficients is obtained by a 'guess' and a continuous improvement of this starting solution by several cycles with improved coefficients.

$$\mathbb{C}_0 \Rightarrow \mathbb{P}_0 \Rightarrow \mathbb{F}_0 \Rightarrow \mathbb{C}_1 \Rightarrow \mathbb{P}_1 \Rightarrow \mathbb{F}_1 \Rightarrow\Rightarrow\Rightarrow \mathbb{C}_n \rightarrow \mathbb{P}_n \rightarrow \mathbb{F}_n.$$

The process converges, if the differences between $\mathbb{P}_n$ and $\mathbb{P}_{n-1}$ have become insignificant. Usually, the limit is set to 10^{-5} for the density matrix elements, which on average corresponds to an energy consistency of $<10^{-6}$ atomic units or $<10^{-3}$ kcal mol^{-1}.

As outlined previously, the density matrix $\mathbb{P}$ modulates the 'field' of the $(n-1)$ other electrons experienced by the one electron under consideration. Therefore, this procedure iteratively to optimise this 'field' until consistency is achieved is also called the *'SCF'* or *'Self Consistent Field Method'*, adding another acronym to the classification of quantitative molecular orbital calculations as the *'LCAO-MO-SCF'* procedure.

The $\mathbb{P}$ matrix is of relevance also in another context, which leads to 'chemically' interesting data that are helpful in the interpretation of a molecule's reactivity. Although the electron distribution can strictly be evaluated only on the basis of $\Psi^*\Psi$, with Ψ being the total probability function, it is desirable for the chemist to assign a partial charge to each atom in the molecule, thus avoiding the cumbersome procedure of evaluating density grids over the whole molecule and to identify the properties of the atoms in the molecular framework. This is accomplished by the so-called *population analysis*, where the electron density is assigned to atoms on the basis of the atomic functions' contributions to the molecular orbitals (i.e., the coefficients). There are several schemes to do this, but the 'classical' one is the *Mulliken* population analysis, based on a summation of the P_{rs} elements, multiplied by the corresponding elements of the overlap integral matrix, S_{rs}. This leads to *'atomic populations'* (and thus partial charges) and *'overlap populations'* between different atoms, and both parameters are frequently utilised in the discussion of the chemistry of a molecule, notably with respect to 'polarisation' and 'bond character'. Although we will deal in more detail with such a usage in the next subsection, at this point it should be emphasised that these data are *not* physical observables and are even basis set-dependent.

5.7
Canonical and Localised Molecular Orbitals and Chemical Model Concepts

The output of a quantitative molecular orbital calculation contains (regulated by the printing options) a large number of data. Most of this information is only needed in special cases or for control purposes, and the chemist would usually concentrate on a selected part of it. The first – and most important – result is the total energy. In case a geometry optimisation has been performed using the gradient method (dependence of the energy on coordinates of the atoms or internal coordinates such as bond lengths and angles), the optimised structure is the next important data, while the energy surface used in the geometry optimisation also supplies frequencies for molecular vibrations useful for comparison with infra-red (IR) and Raman spectra.

The composition of the eigenvectors and the one-electron energies associated with them is also of some interest, but here we leave the field of physical observables. As mentioned in connection with atomic populations, only the total energy and the total density $\Psi^*\Psi$ are of physical significance, and thus the eigenvectors (i.e., the molecular orbitals) can be changed by any unitary transformation to completely different compositions and thus shapes. We will shortly make use of this possibility to discuss another very common model in chemistry.

The eigenvectors resulting from a Hartree–Fock LCAO-MO-SCF calculation are called *'canonical MOs'*, and the associated pseudo-eigenvalues, the 'orbital energies', are associated with measurable quantities by *Koopmans' theorem*, which states that these energies should correspond to the vertical ionisation energies. Vertical in this context means that the system being ionised would not react fast enough to the new electronic situation by relaxation and hence, the leaving electron would have a kinetic energy which corresponds to the difference between excitation energy and the energy level that the electron had occupied in the molecule. Koopmans' theorem is used to classify ESCA (Electron Spectroscopy for Chemical Analysis) and photoelectron (PE) spectra, where this kinetic energy is measured, by assigning the spectral lines to the molecular orbitals and their respective one-electron energies. The assumption of a system relaxing so slowly that the leaving electron would not 'feel' it, might be reasonable for nuclear relaxation, but not so reasonable for the adaptation of the $(n-1)$ rest electrons, and this is the reason why one speaks of a *'Koopmans' defect'*, whenever the theorem apparently fails. Therefore, it is also inappropriate to cite ESCA and PE spectra as 'proof' of the 'real existence' of molecular orbitals. We know that they are not observable quantities, but rather a product of our inability to solve the Schrödinger equation for an n-electron system and the introduced approximation using one-electron functions. The validity of any model based on the canonical MOs and the one-particle energies is very limited, therefore, to some aspects of the quantum theoretical results for a chemical system.

We have stated that any type of unitary transformation of the molecular orbitals leads to an equally 'true' or – more exactly – an equally valid solution. The most relevant transformations of this type are being performed with the purpose of restricting the participation of atomic orbitals (= basis functions) in each molecular orbital to the lowest possible number of atoms. This procedure is known as *localisation of molecular orbitals*, and it can be accomplished according to criteria introduced by *Edmiston* and *Rüdenberg* (maximisation of self-repulsion $\langle ii|\frac{1}{r_{12}}ii\rangle$) and *Boys* (maximisation of the dipole moment in bond direction $\langle ii|\frac{1}{r_{12}^2}ii\rangle$), where in both cases i stands for spin orbitals $|\psi_i\rangle$.

After performing such a transformation we obtain representations of the electron distribution in a molecule, which correspond very well to the common *Lewis* formulae, where electrons are shown as one or more lines connecting atoms or as 'lone pairs' not contributing to a bond. This picture, however, is now obtained on the basis of a quantitative molecular orbital calculation and can thus be regarded as a 'quantification' of the Lewis formulae. In contrast, one can regard the Lewis formulae, therefore, as a really ingenious zeroth order approximation to the quantum theoretical representation of molecules, and there is much reason to regard these formulae as a superior model for a qualitative dicussion of chemistry than all the mix of qualitative VB plus MO pictures. If we add the atomic populations as an approximative quantitative description of the partial charges δ^+ and δ^- usu-

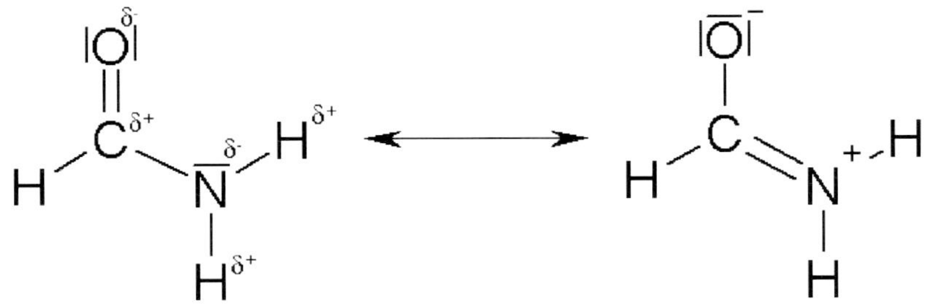

Fig. 5.1 The 'classical' Lewis formula of formamide.

ally written in Lewis formulae, and look more closely at the localised MOs analysing percentual contributions of atomic functions to bond orbitals and/or density plots, we obtain an excellent qualitative or even semiquantitative image of chemical properties, based on an *ab initio* quantum theoretical treatment. As today such calculations can be performed for smaller and medium-sized molecules on a personal computer within a few minutes, there is no reason to stick to one-electron models or even some obscure qualitative orbital pictures, which are certainly much farther away from exact theory (and probably also from reality!) than the 'old-fashioned' Lewis formulae.

We will exemplify this with the quantitative treatment of formamide, a simple but in many aspects representative, molecule. Formamide ($HCONH_2$) shows some interesting properties, as the rotation around the C–N bond is hindered, leading to two different ^1H signals for the N–H hydrogens in the proton NMR spectrum, which can be brought to coalescence, however, at elevated temperatures. This observation has been attributed to a partial double bond character of the C–N bond, and a classical Lewis formula of the compound illustrates this, as shown in Fig. 5.1.

A quantitative *ab initio* MO treatment of the molecule with a simple basis set of GTOs listed in Table 5.1, yields the canonical MOs listed in Table 5.2.

One can see that almost every valence MO consists of contributions of several atoms, and thus covers the whole molecular space. After the localisation procedure, the MOs listed in Table 5.3 are obtained, and from their composition one immediately sees that they are concentrated on just two atoms – that is, to one bond, or even, in the case of oxygen, to one atom alone, thus representing 'lone pairs'. Figure 5.2 shows the density plots of the various localised MOs, and the H–C and N–H bonds and two lone pairs at the O atom can be clearly recognised. Further, two bonds are seen for C=O and, most interestingly, there is no distinct lone pair at nitrogen, but rather two C–N bonds, both with a centre of density strongly shifted towards nitrogen. This clearly illustrates a weak, but significant 'double bond character', which is surely weaker than in the case of the C–O bond, where the centre of density is much closer to the middle of the bonds.

The atomic populations according to Mulliken have been listed at the respective atoms, and they fully confirm and quantify the partial charge distributions expected intuitively by the chemist. They even explain the different NMR signals

Table 5.1 STO-6G single-zeta basis set used for LCAO-MO-SCF model calculations of formamide. Each Slater-type orbital is represented by six Gauss-type orbitals (GTOs) of the formula $\varphi_{GTO} = N \cdot e^{-\alpha \cdot r^2} \cdot Y^{l,m}_{\theta,\varphi}$; the exponents α of the Gaussians are listed in the table.

Atom	STO	GTO_1	GTO_2	GTO_3	GTO_4	GTO_5	GTO_6
C	1s	0.0	0.0	100.0	2.2	0.3	3.1
C	2s	0.0	0.0	0.0	12.4	2.4	33.5
C	2px	0.0	0.0	0.0	2.7	4.1	4.7
C	2py	0.0	0.0	0.0	0.2	0.7	3.2
C	2pz	0.0	0.0	0.0	0.0	0.0	0.0
O	1s	100.0	0.0	0.0	5.3	0.7	0.7
O	2s	0.0	0.0	0.0	70.2	10.3	14.2
O	2px	0.0	0.0	0.0	3.6	0.0	1.7
O	2py	0.0	0.0	0.0	0.1	0.1	0.7
O	2pz	0.0	0.0	0.0	0.0	0.0	0.0
N	1s	0.0	100.0	0.0	0.3	5.8	0.3
N	2s	0.0	0.0	0.0	2.7	69.9	5.0
N	2px	0.0	0.0	0.0	0.1	0.0	2.7
N	2py	0.0	0.0	0.0	0.1	0.1	11.1
N	2pz	0.0	0.0	0.0	0.0	0.0	0.0
H_1	1s	0.0	0.0	0.0	0.2	0.1	11.8
H_2	1s	0.0	0.0	0.0	0.0	2.6	5.7
H_3	1s	0.0	0.0	0.0	0.0	3.0	1.5

Table 5.2 Percentual composition of canonical molecular orbitals for formamide, calculated with single-zeta basis set

Atom	AO	1	2	3	4	5	6
C	1s	0.0	0.0	100.0	2.2	0.3	3.1
C	2s	0.0	0.0	0.0	12.4	2.4	33.5
C	2px	0.0	0.0	0.0	2.7	4.1	4.7
C	2py	0.0	0.0	0.0	0.2	0.7	3.2
C	2pz	0.0	0.0	0.0	0.0	0.0	0.0
O	1s	100.0	0.0	0.0	5.3	0.7	0.7
O	2s	0.0	0.0	0.0	70.2	10.3	14.2
O	2px	0.0	0.0	0.0	3.6	0.0	1.7
O	2py	0.0	0.0	0.0	0.1	0.1	0.7
O	2pz	0.0	0.0	0.0	0.0	0.0	0.0

Table 5.2 *(continued)*

Atom	AO	1	2	3	4	5	6
N	1s	0.0	100.0	0.0	0.3	5.8	0.3
N	2s	0.0	0.0	0.0	2.7	69.9	5.0
N	2px	0.0	0.0	0.0	0.1	0.0	2.7
N	2py	0.0	0.0	0.0	0.1	0.1	11.1
N	2pz	0.0	0.0	0.0	0.0	0.0	0.0
H_1	1s	0.0	0.0	0.0	0.2	0.1	11.8
H_2	1s	0.0	0.0	0.0	0.0	2.6	5.7
H_3	1s	0.0	0.0	0.0	0.0	3.0	1.5

Atom	AO	7	8	9	10	11	12
C	1s	0.2	0.1	0.2	0.0	0.0	0.0
C	2s	1.9	1.2	2.2	0.0	0.1	0.0
C	2px	0.0	3.6	15.8	0.0	0.1	0.0
C	2py	16.1	16.3	2.3	0.0	0.5	0.0
C	2pz	0.0	0.0	0.0	38.3	0.0	4.7
O	1s	0.2	0.2	0.7	0.0	0.0	0.0
O	2s	4.9	6.3	24.2	0.0	0.0	0.0
O	2px	3.2	2.0	40.2	0.0	1.8	0.0
O	2py	5.7	11.0	0.2	0.0	73.6	0.0
O	2pz	0.0	0.0	0.0	26.9	0.0	33.1
N	1s	0.0	0.0	0.0	0.0	0.1	0.0
N	2s	0.1	0.1	0.1	0.0	4.2	0.0
N	2px	38.3	0.3	4.7	0.0	3.5	0.0
N	2py	1.2	31.0	0.3	0.0	3.8	0.0
N	2pz	0.0	0.0	0.0	34.8	0.0	62.3
H_1	1s	3.2	7.0	3.7	0.0	10.6	0.0
H_2	1s	3.2	18.3	1.5	0.0	0.9	0.0
H_3	1s	22.1	2.7	4.0	0.0	0.7	0.0

Table 5.3 Percentual composition of localised molecular orbitals for formamide, calculated with single-zeta basis set

Atom	AO	1	2	3	4	5	6
C	1s	0.0	0.0	100.0	0.6	0.4	1.2
C	2s	0.0	0.0	0.0	8.5	5.6	20.1
C	2px	0.0	0.0	0.0	9.0	4.1	5.4
C	2py	0.0	0.0	0.0	0.3	4.1	25.7
C	2pz	0.0	0.0	0.0	20.5	2.4	0.0
O	1s	99.4	0.0	0.0	0.4	0.0	0.0
O	2s	0.5	0.0	0.0	5.6	0.0	0.3
O	2px	0.1	0.0	0.0	17.7	0.2	0.3
O	2py	0.0	0.0	0.0	0.4	0.3	1.6
O	2pz	0.0	0.0	0.0	35.4	1.0	0.0
N	1s	0.0	99.9	0.0	0.0	0.5	0.0
N	2s	0.0	0.1	0.0	0.1	14.0	1.8
N	2px	0.0	0.0	0.0	0.1	6.1	0.3
N	2py	0.0	0.0	0.0	0.0	6.5	0.5
N	2pz	0.0	0.0	0.0	1.3	53.3	0.0
H_1	1s	0.0	0.0	0.0	0.1	0.4	42.4
H_2	1s	0.0	0.0	0.0	0.0	0.4	0.5
H_3	1s	0.0	0.0	0.0	0.1	0.4	0.1

Atom	AO	7	8	9	10	11	12
C	1s	0.1	0.0	0.0	0.6	0.0	0.4
C	2s	1.3	1.0	1.5	8.5	1.4	5.6
C	2px	0.4	0.5	0.4	9.0	1.8	4.1
C	2py	0.4	0.5	2.1	0.3	1.1	4.1
C	2pz	0.0	0.0	0.0	20.5	0.0	2.4
O	1s	0.0	0.0	0.8	0.4	0.9	0.0
O	2s	0.1	0.1	41.9	5.6	42.6	0.0
O	2px	0.2	0.0	16.4	17.7	3.8	0.2
O	2py	0.1	0.0	34.4	0.4	46.5	0.3
O	2pz	0.0	0.0	0.0	35.4	0.0	1.0
N	1s	1.2	1.2	0.0	0.0	0.0	0.5
N	2s	24.2	24.1	1.1	0.1	0.6	14.0
N	2px	35.5	2.0	0.2	0.1	0.1	6.1
N	2py	1.6	35.8	0.2	0.0	0.1	6.5
N	2pz	0.0	0.0	0.0	1.3	0.0	53.3

Table 5.3 *(continued)*

Atom	AO	1	2	3	4	5	6
H_1	1s	0.0	0.3	0.5	0.1	1.1	0.4
H_2	1s	1.2	33.2	0.0	0.0	0.0	0.4
H_3	1s	33.6	1.3	0.1	0.1	0.0	0.4

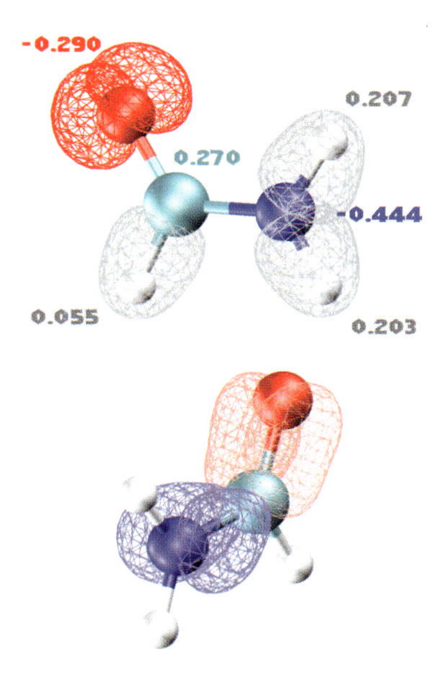

Fig. 5.2 Localised molecular orbitals of formamide and partial charges of the atoms according to Mulliken population analysis.

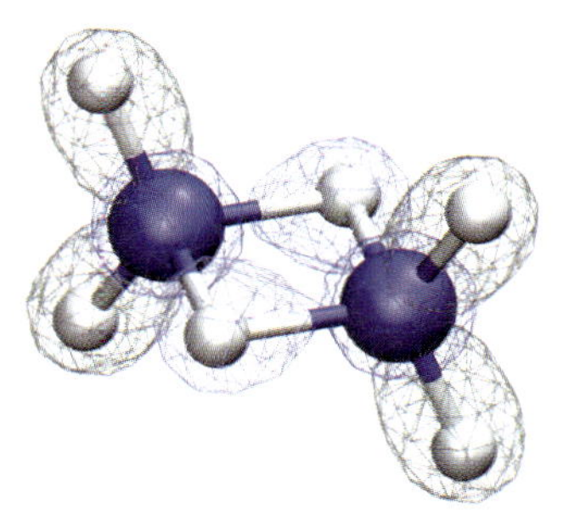

Fig. 5.3 Localised molecular orbitals for diborane.

for the N–H protons, and that the H–C signal is found at higher field than the N–H signals, due to a higher electron density at H(C) and thus a better shielding of the proton than for either H(N) (Fig. 5.2).

There are examples, however, where the localisation procedure does not yield just lone pairs and bonds between two atoms. Diborane is such an example, and its localised MOs are depicted in Figure 5.3. Here, there are two MOs extending over three atoms, namely the $B \cdots H \cdots B$ bridges. This confirms the classical Lewis formula with two 3-center/2-electron bonds.

Another example for such 'delocalised' MOs or bonds are found in benzene and other aromatic compounds, where the organic chemists like to talk about delocalised 'π electrons' (later we will see, why they are *not* π electrons). Thus, the quantitative MO theory finally seems to bring us back to a renaissance of the 'classical' – but nevertheless more appropriate – models of chemistry.

Despite this pleasant and convenient property of localised MOs, one should never forget that they are also *not* physical observables, but just one possibility to decompose the total probability function into one-particle functions, and hence any model derived on this basis will again only reflect some of the aspects of the system under consideration.

Localised MOs, however, are not only of major importance for the visualisation of quantitative results of quantum chemical calculations, but they are also utilised in sophisticated calculation methods exceeding the accuracy of the Hartree–Fock method, and for this reason we will encounter them again in Chapter 8.

Test Questions Related to this Chapter

1. Which errors could occur due to a non-relativistic treatment of a molecule?
2. What consequences has the adiabatic approximation for the evaluation of IR and Raman spectra?
3. What is the reason for the energy term 'electron correlation'?
4. What is the physical meaning of spin orbitals and orbital energies?
5. How can we classify 'atomic orbitals' and 'molecular orbitals' in terms of vector space theory of matter?
6. How is spin introduced in Slater determinants, and how do we classify them according to their total spin?
7. How do we optimise Slater determinants, and what is the criterion for an optimal function within a given basis set?
8. What criteria do we have to test the quality of basis sets?
9. Why can we change the composition of the spin orbitals by unitary transformations, although this may completely change their shape?
10. What model notations of molecules are suitable for depicting different aspects of chemical behaviour, according to a quantitative quantum mechanical treatment?

6
Perturbation Theory in Quantum Chemistry

He who can properly define and divide is to be considered a god.

Plato, 4th century BC

This chapter deals with an important methodology that has been used successfully in theoretical physics for more than a century, and also forms the basis of well-established procedures in theoretical chemistry. The principal idea is the consideration of additional or previously neglected effects as a 'perturbation' of an 'unperturbed' system, which has been calculated previously, and to determine the influence of this perturbation on the eigenvalues and eigenfunctions of the operator used in the original calculation.

6.1
Projections and Projectors

In order to discuss the perturbational approach, the terminology of 'projecting' one or more functions to another function and the associated '*projection operators*' (shortly called '*projectors*') will be introduced. This will be done using the already well-known example of the linear combination of basis functions φ_i to construct a more complicated function Ψ:

$$|\Psi\rangle = c_1|\varphi_1\rangle + c_2|\varphi_2\rangle + \cdots + c_n|\varphi_n\rangle = \sum_i c_i|\varphi_i\rangle$$

The scalar product of $|\Psi\rangle$ with itself is given by

$$\langle\Psi|\Psi\rangle = c_1^2|\langle\varphi_1|\varphi_1\rangle + c_2^2\langle\varphi_2|\varphi_2\rangle + \cdots + c_n^2\langle\varphi_n|\varphi_n\rangle = \sum_i c_i^2\langle\varphi_i|\varphi_i\rangle$$

If the basis $\{\varphi_i\}$ consists of orthonormalised functions, and if the function $|\Psi\rangle$ itself is normalised, i.e., $\langle\Psi|\Psi\rangle = 1$, this leads to $\sum_i c_i^2 = 1$.

The Basics of Theoretical and Computational Chemistry. Edited by B. M. Rode, T. S. Hofer and M. D. Kugler

ISBN: 978-3-527-31773-8

If we form the scalar product of any of the basis functions φ_i with the total function Ψ, we obtain the linear combination coefficient and thus the contribution of $|\varphi_i\rangle$ to the function $|\Psi\rangle$

$$\langle\varphi_i|\Psi\rangle = c_i$$

the squared value of which, c_i^2, corresponds to the probability of φ_i to occur in the total probability $\langle\Psi|\Psi\rangle$. The scalar product $\langle\varphi_i|\Psi\rangle$ is also called the *projection* of $|\varphi_i\rangle$ onto $|\Psi\rangle$.

Thus, we can reformulate the summation over all n basis function as

$$\sum_i^n \langle\varphi_i|\Psi\rangle|\varphi_i\rangle = |\Psi\rangle \quad \text{or} \quad \sum_i^n |\varphi_i\rangle\langle\varphi_i|\Psi\rangle = |\Psi\rangle$$

As the expression $\sum_i^n |\varphi_i\rangle\langle\varphi_i|$ corresponds to the identity operator $\hat{1}$, this expression is also known as the *resolution of the identity.* If the summation refers only to a part of the basis functions $m < n$, i.e., to a subspace of the vector space spanned by the n basis functions, this operator is called a *projector*, as it 'projects' only the contributions of the functions up to $|\varphi_m\rangle$.

$$\sum_i^n |\varphi_i\rangle\langle\varphi_i| = \hat{1} \quad \sum_i^{m<n} |\varphi_i\rangle\langle\varphi_i| = \hat{P}_m$$

For every projector $\hat{P}_m$ a complementary projector for the rest of basis functions exists, defined by $[\hat{1} - \hat{P}_m]$.

For expectation values of operators, we can equally perform a resolution. Let us assume $|\Psi\rangle$ to be a normalised function, thus simplifying the expression for the expectation value of an operator $\hat{O}$ to

$$\langle\hat{O}\rangle = \langle\Psi|\hat{O}\Psi\rangle = \sum_i^n \langle\varphi_i c_i|\hat{O}\varphi_i c_i\rangle = \sum_i^n c_i^2\langle\varphi_i|\hat{O}\varphi_i\rangle = \sum_i^n c_i^2 o_i$$

where o_i are the eigenvalues associated with the operator $\hat{O}$, and $c_i^2 = \langle\varphi_i|\Psi\rangle^2$ correspond to the probabilities of these eigenvalues. We can make use of these 'projections' of $|\varphi_i\rangle$ onto the function $|\Psi\rangle$ to rewrite the last part of the previous expression as

$$\langle\hat{O}\rangle = \sum_i^n c_i^2 o_i = \sum_i^n o_i\langle\varphi_i|\Psi\rangle^2 = \sum_i^n \langle\Psi|\varphi_i\rangle o_i\langle\varphi_i|\Psi\rangle$$

A comparison of the last expression with the common formulation of the expectation value $\langle\Psi|\hat{O}\Psi\rangle$ shows that we have obtained a new form to express the operator $\hat{O}$ as

$$\hat{O} = \sum_i |\varphi_i\rangle o_i \langle\varphi_i|$$

which is called the *spectral form of the operator* $\hat{O}$. We can again formulate projectors which only refer to a subset of eigenvalues o_i, $\hat{P}_m$

$$\hat{P}_m = \sum_i^{m<n} |\varphi_i\rangle o_i \langle\varphi_i|$$

All kinds of projection operators have specific properties:

1. They are self-adjugate, i.e. $\hat{P} = \hat{P}^+$.
2. They are idempotent, i.e., $\hat{P} = \hat{P}^2 = \hat{P}^3 \cdots = \hat{P}^n$.
3. They are mutually exclusive (complementary).
4. The eigenvalue of a projector is either 1 or 0. The associated variable is, therefore, called the 'variable of existence'.

We will utilise projections and projectors in perturbation theory, and again in Chapter 7 when dealing with group theory.

6.2 Principles of Perturbation Theory

We will discuss the perturbational methods by using their most common case in theoretical chemistry, which is its application to the Hamiltonian. The unperturbed Hamilton operator $\hat{H}^\circ$ delivers the energy value E° and the unperturbed eigenfunction $|\Phi^\circ\rangle$. The 'perturbation' can be an additional external influence such as an electric or magnetic field, but it could also be any kind of correction we want to perform due to previously neglected contributions (e.g., electron correlation). This additional effect is represented by a 'perturbation operator' $\hat{V}$, which corresponds to the physical nature of the 'perturbation' and is added to the unperturbed Hamiltonian with a scalar scaling parameter λ; this allows the amount of the 'perturbation' to be modulated, thus giving the definition of the total Hamiltonian

$$\hat{H} = \hat{H}^\circ + \lambda \cdot \hat{V}$$

The eigenfunction of the unperturbed Hamiltonian $|\Phi^\circ\rangle$ will contribute to the total probability function with coefficient 1, according to

$$\langle\Phi^\circ|\Psi\rangle = 1$$

which is called '*intermediate normalisation*'. The energy and the probability function can be developed as series of λ^n terms as

$$E = E^\circ + \lambda\varepsilon_1 + \lambda^2\varepsilon_2 + \cdots + \lambda^n\varepsilon_n$$
$$|\Psi\rangle = |\phi^\circ\rangle + \lambda|\phi^1\rangle + \lambda^2|\phi^2\rangle + \cdots + \lambda^n|\phi^n\rangle$$

The energies ε_i are called *perturbation energies* of the ith order, and the corresponding functions ϕ^i are the *perturbation functions* of the ith order.

When one substitutes these expressions in the Schrödinger equation $\hat{H}|\psi\rangle = E|\psi\rangle$, one obtains

$$\left(H^\circ + \lambda\hat{V} - E^\circ - \sum_i \lambda^i\varepsilon_i\right)\left(|\phi^\circ\rangle + \sum_i \lambda^i|\phi^i\rangle\right) = \emptyset$$

After performing the multiplications of this expression it is helpful to order and collect the terms according to the powers of λ, which leads to the expression

$$H^\circ|\phi^\circ\rangle - E^\circ|\phi^\circ\rangle + \lambda[(H^\circ - E^\circ)|\phi^1\rangle + \hat{V}|\phi^\circ\rangle - \varepsilon_1|\phi^\circ\rangle] +$$
$$+ \lambda^2[(H^\circ - E^\circ]|\phi^2\rangle - \varepsilon_2|\phi^\circ\rangle + \hat{V}|\phi^1\rangle - \varepsilon_1|\phi^1\rangle] + \cdots = \emptyset$$

As this expression must be valid for any arbitrary value of λ, all expressions inside the brackets [...] must be zero. This leads to separate equations of increasing order, and the unperturbed Schrödinger equation represents the *zero*th order.

$$\begin{aligned}
&(0) \quad (\hat{H}^\circ - E^\circ)|\phi^\circ\rangle = \emptyset \\
&(1) \quad (\hat{H}^\circ - E^\circ)|\phi^1\rangle + \hat{V}|\phi^\circ\rangle - \varepsilon_1|\phi^\circ\rangle = \emptyset \\
&(2) \quad (\hat{H}^\circ - E^\circ)|\phi^2\rangle + \hat{V}|\phi^1\rangle - \varepsilon_1|\phi^1\rangle - \varepsilon_2|\phi^\circ\rangle = \emptyset \\
&\qquad\qquad \Downarrow \\
&(n) \quad (\hat{H}^\circ - E^\circ)|\phi^n\rangle + \hat{V}|\phi^{n-1}\rangle - \sum_{j=0}^{n} \varepsilon_{n-j}|\phi^j\rangle = \emptyset
\end{aligned}$$

Having in mind the intermediate normalisation $\langle\phi^\circ|\psi\rangle = 1$, which implies $\langle\phi^\circ\ \phi^n\rangle = \delta_{\emptyset,n}$, we will now project the unperturbed vector $|\phi^\circ\rangle$ onto the perturbation equations. For the first order this gives

$$\langle\phi^\circ|H^\circ - E^\circ|\phi^1\rangle + \langle\phi^\circ|\hat{V}|\phi^\circ\rangle = \varepsilon_1 \underbrace{\langle\Phi^\circ|\Phi^\circ\rangle}_{=1}$$

In the first term, the Hamiltonian acts as Hermitian operator on ϕ° in the bra space, as it acts in the ket space (i.e., producing E°), and thus the operator $H^\circ - E^\circ$ annihilates this term. What remains, is

$$\varepsilon_1 = \langle \phi^\circ | \hat{V} | \phi^\circ \rangle$$

the expression for the perturbation energy of the first order. For the nth order the projection yields

$$\langle \phi^\circ | \hat{H}^\circ - E^\circ | \phi^n \rangle + \langle \phi^\circ | \hat{V} | \phi^{n-1} \rangle - \sum_{j=0}^{n} \varepsilon_{n-j} \langle \phi^\circ | \phi^j \rangle$$

The first term is again annihilated as in the first-order case, and due to the intermediate normalisation all terms in the sum become zero, except that for j = $\emptyset$. Thus, the resulting expression gives the formula for the perturbation energy of the nth order

$$\langle \phi^\circ | \hat{V} | \phi^{n-1} \rangle = \varepsilon_n$$

We see that it is possible to calculate the perturbation energy of the nth order from the function of the $(n-1)^{st}$ order.

For calculations of perturbation energies of an order > 1, perturbation functions are needed. There are several procedures to construct such functions, and one of the most common and prominent will be presented in the next section.

6.3 The Rayleigh–Schrödinger Perturbation Method

In this approach the perturbation functions $|\phi^n\rangle$ are constructed as linear combinations of the eigenfunctions $|\phi_i^\circ\rangle$ of the unperturbed Hamiltonian other than the ground state function $|\phi^\circ\rangle$. That means that the eigenfunctions of excited states serve as a basis set for the construction of the perturbation functions. This does not necessarily mean performing a large number of calculations for these states by solving the Schrödinger equation: from the ground state calculations one obtains a number of 'virtual' molecular orbitals as a byproduct. Although these virtual orbitals have no physical meaning, they can be utilised to construct (pseudo-) excited spin determinants, where one or more electrons are placed into molecular orbitals with higher energy values instead of their ground state functions. Although not being excited states in the sense of physics, they are a convenient tool to construct a sufficiently good mathematical basis for the purpose of developing perturbation functions. The same construction of 'excited' configurations is also used in multideterminantal methods to calculate electron correlation, which will be discussed in Chapter 8.

The Rayleigh–Schrödinger construction of the perturbation functions can be written as

$$|\phi^n\rangle = \sum_i c_i^{(n)} |\phi_i^*\rangle \quad \text{or} \quad |\phi^n\rangle = \sum_i |\phi_i^*\rangle\langle\phi_i^*|\phi^n\rangle$$

where $|\phi_i^*\rangle$ are eigenfunctions of the unperturbed Hamiltonian, characterised by an asterisk to indicate their correspondence to 'excited' states. The coefficients $c_i^{(n)}$ are the linear combination coefficients for the nth order perturbation function.

We will now project the functions $|\phi_i^*\rangle$ onto the perturbation equation of the nth order; this leads to

$$\langle\phi_i^*|\hat{H}^\circ - E^\circ|\phi^n\rangle + \langle\phi_i^*|\hat{V}|\phi^{n-1}\rangle - \sum_{j=0}^{n} \varepsilon_{n-j}\langle\phi_i^*|\phi^j\rangle = \emptyset$$

which can be rearranged to

$$\langle(\hat{H}^\circ - E^\circ)\phi_i^*|\phi^n\rangle + \langle\phi_i^*|\hat{V}|\phi^{n-1}\rangle - \sum_{j=1}^{n} \varepsilon_{n-j}\langle\phi_i^*|\phi^j\rangle = \emptyset$$

The summation starts with $j = 1$ now, as the functions $|\phi_i^*\rangle$ and $|\phi^\circ\rangle$ are orthogonal, being both eigenfunctions of the same Hermitian operator $\hat{H}^\circ$ and making all elements $\langle\phi_i^*|\phi^\circ\rangle$ vanish. In the first part of the expression the operator $\hat{H}^\circ$ acting on $|\phi_i^*\rangle$ produces E_i°, the eigenvalue associated with the function $|\phi_i^*\rangle$. Thus, we can formulate

$$(E_i^\circ - E^\circ)\underbrace{\langle\phi_i^*|\phi^n\rangle}_{=c_i^{(n)}} = -\langle\phi_i^*|\hat{V}|\phi^{n-1}\rangle + \sum_{j=1}^{n} \varepsilon_{n-j}\langle\phi_i^*|\phi^j\rangle$$

and hence obtain the final expression for the linear combination coefficients for any of the desired perturbation functions as

$$c_i^{(n)} = \frac{1}{E^\circ - E_i^\circ}\left[\langle\phi_i^*|\hat{V}|\phi^{n-1}\rangle - \sum_{j=1}^{n} \varepsilon_{n-j}\langle\phi_i^*|\phi^j\rangle\right]$$

We will make use of this approach now to construct the first-order perturbation function, which will enable us to calculate the perturbation energy up to the second order. The coefficients for $|\phi^1\rangle$ result as

$$c_i^{(1)} = \frac{1}{E^\circ - E_i^\circ}\langle\phi_i^*|\hat{V}|\phi^\circ\rangle$$

as the second term with $\varepsilon_{n-j} = \varepsilon_\circ$ vanishes. With these coefficients the expression for the first-order perturbation function becomes

$$|\phi^1\rangle = \sum_i c_i^{(1)}|\phi_i^*\rangle = \sum_i \frac{\langle\phi_i^*|\hat{V}|\phi^\circ\rangle}{E^\circ - E_i^\circ}|\phi_i^*\rangle$$

and the total function for the perturbed system results as

$$|\psi\rangle = |\phi^\circ\rangle + \sum_i \frac{\langle \phi_i^* | \hat{V} | \phi^\circ \rangle}{E^\circ - E_i^\circ} |\phi_i^*\rangle$$

The total energy of the perturbed system up to the second order is then obtained as

$$E = E^\circ + \varepsilon_1 + \varepsilon_2 = E^\circ + \langle \phi^\circ | \hat{V} | \phi^\circ \rangle + \sum_i \frac{\langle \phi^\circ | \hat{V} | \phi_i^* \rangle^2}{E^\circ - E_i^\circ}$$

In the same way (but with more complicated terms appearing), one can construct perturbation functions of higher order.

As previously indicated, there are other ways to formulate perturbation theory and to construct perturbation functions, but the example given appears sufficient for demonstrating this methodical approach. It has still to be discussed, however, in which cases a perturbational approach seems adequate, and what the potential problems of this method might be.

6.4 Applications of Perturbation Theory in Quantum Chemistry

A general guideline for the successful application of a perturbational approach is the relationship between unperturbed energy and the perturbational contribution. The latter should never come to the same order of magnitude, but ideally should be one order smaller, as otherwise the results could deviate strongly from 'reality'.

A second important point to consider is a possible oscillating behaviour of the perturbation contributions of increasing order, which might even be diverging. If no convergence is observed, the perturbational approach might not be appropriate.

When performing perturbational calculations one should also be aware of the possibility that this method cannot only underestimate, but also overestimate, the effects.

The most typical applications of perturbational calculations deal with real perturbations of a system by external influence, such as the *Zemann effect*, which describes the change of electronic levels and thus the observed spectral lines by a magnetic field. The analogous effect of an electric field is known as the *Stark effect*, and is equally accessible via perturbational calculations. In both cases, the parameter λ preceding the perturbation operator is a convenient instrument to vary the strength of the field.

As mentioned above, the posterior introduction of previously neglected terms or contributions in quantum chemical calculations is another important application of perturbational methods. The most widely use of this approach is the correction for electron correlation effects, as practised in particular in the Møller–Plesset MP/n methods, which will be discussed in Chapter 8.

Test Questions Related to this Chapter

1. How can we modulate the amount of perturbation?
2. What conditions should be fulfilled for the successful application of a perturbational approach?
3. What are the main error sources in perturbational calculations?
4. What is the principle of the Rayleigh–Schrödinger perturbation method?
5. What simplified approach can be used to obtain a set of eigenfunctions for the unperturbed Hamiltonian representing electronic configurations other than the ground state?

7
Group Theory in Theoretical Chemistry

The mathematical sciences particularly exhibit order, symmetry and limitation; and these are the greatest forms of the beautiful.

Aristoteles, 4th century BC

Mathematical group theory is a very universal and versatile tool in physics and chemistry, in particular in connection with the symmetry of systems. Group theory is used equally by quantum chemistry and spectroscopy alike, and it frequently occurs that the chemist's common language unknowingly contains group theory nomenclature. Therefore, this chapter will outline the basics of group theory before focusing on specific chemical applications.

7.1
Definition of a Group

In mathematical terms, a group is a structured set with a unique algebraic relationship between all of its elements. This relationship – often called 'group multiplication' (although not necessarily a multiplication!) – will be annotated, wherever specifically required, by the small circle '$\circ$'. In order to form a group G, the structured set must fulfil the following conditions, also called *group postulates*:

1. *Closure* under a unique algebraic relationship:

   $$a \circ b = c \in G \; \forall a, b \in G$$

   Every element of the group combined with another element of it leads to a third element belonging to the same group, and the group is thus closed under the algebraic operation.
2. *Associativity* of group elements:

   $$(a \circ b) \circ c = a \circ (b \circ c)$$

The Basics of Theoretical and Computational Chemistry. Edited by B. M. Rode, T. S. Hofer and M. D. Kugler

ISBN: 978-3-527-31773-8

3. *Identity element*: There is an element E in the group, which combined with any other element a yields the same, unchanged element a:

$$a \circ E = a \ \forall a \in G$$

This element E is called the identity element.

4. *Inverse elements*: For every element a an inverse element a^{-1} must exist in the group which, upon combination with a, yields the identity element:

$$a \circ a^{-1} = E \ \forall a, a^{-1} \in G$$

There are a few theorems concerning the group elements:
- there is a left inverse element, which is identical with the right inverse element: $a \circ a^{-1}_{right} = a^{-1}_{left} \circ a$; $a^{-1}_{right} \equiv a^{-1}_{left}$;
- there is a left identity element, which is identical with the right identity element: $a \circ E_{right} = E_{left} \circ a$; $E_{right} \equiv E_{left}$;
- the inverse of an element is always unique; and
- the identity element of a group is unique.

We can illustrate the conditions to form a group by examining the validity of the postulates with a few examples:
- all integers under addition ($\circ \equiv +$) form a group;
- all rational numbers under multiplication form a group iff $\emptyset$ is excluded; and
- all integers under subtraction ($\circ \equiv -$) do not form a group, because of the lack of associativity: $(4 - 2) - 1 \neq 4 - (2 - 1)$.

One can summarise all elements of a group and show their inter-relationships by a so-called *multiplication table*. An example of such a multiplication table is given in Table 7.1 for the permutation group S_3, which consists of six elements, the notations and operations of which are illustrated in the upper part of the table. The combination of any element a of a row m with any element b of a column n delivers the result of $a \circ b$ as element a_{mn} of the table. No element can occur twice in either a row or a column, and every row and column of the multiplication table contains all group elements.

The multiplication table is also convenient to illustrate a few further definitions of group theory. One of these is the *subgroup* $H \subset G$, which contains the identity element and a subset of the group elements, forming thereby a group of a smaller dimension (the group G itself and the identity element alone are 'improper' subgroups). In our example, the identity element E and the element (12) form the subgroup S_2.

The *cyclic groups* are a special case, as they consist of a 'generator' g and its multiples. For a group of order n, g^n corresponds to the identity element:

Table 7.1 Multiplication table for permutation group S_3.

Permutation operators (elements of the group):	**Classes**
(1)(2)(3) ABC = ABC (Identity Operator E)	E
(12) ABC = BAC	(12), (13), (23)
(13) ABC = CBA	(123), (132)
(23) ABC = ACB	
(123) ABC = CAB	
(132) ABC = BCA	

Multiplication table

S_3	**E**	**(12)**	**(13)**	**(23)**	**(123)**	**(132)**
E	E	(12)	(13)	(23)	(123)	(132)
(12)	(12)	E	(123)	(132)	(13)	(23)
(13)	(13)	(132)	E	(123)	(23)	(12)
(23)	(23)	(123)	(132)	E	(12)	(13)
(123)	(123)	(23)	(12)	(13)	(132)	E
(132)	(132)	(13)	(23)	(12)	E	(123)

Inverse Elements: $E^{-1} = E$, $(12)^{-1} = (12)$, $(13)^{-1} = (13)$, $(23)^{-1} = (23)$
$(123)^{-1} = (132)$, $(132)^{-1} = (123)$

$$G = \{g^1, g^2 \ldots g^n \equiv E\}$$

We will encounter such a group later among the symmetry groups, namely the group C_3.

Abelian groups are defined by:

$$a \circ b = b \circ a \quad \forall a, b \in G,$$

that is, the group elements are commutative. As an example we will later see the symmetry group C_{2v}.

There are two types of mapping of groups: (i) those where a bidirectional unique relationship exists between the pre-image and image groups, called *isomorphism*; and (ii) those where the mapping relationship is only unidirectionally unique, from pre-image to image group, called *homomorphism*. In the latter type, more than one element of the pre-image group can be mapped onto one element of the image group. The elements of the pre-image group mapped onto the identity element of the image group, are called the '*kernel*' of the homomorphism, while the number of pre-image group elements mapped onto one element of the image group is termed 'multiplicity' of the homomorphism, or the 'order of the kernel'.

The famous *Cayley theorem* discloses a very important finding of group theory, stating that every finite group of order n is isomorphic to a subgroup of the permutation group S_n (which is of the order $n!$). This is the reason why the first example in this chapter was chosen from the permutation groups.

The next important definition concerns *classes* of elements within a group. Let us assume that a group element c is formed from another element a by

$$c = b^{-1} \circ a \circ b \quad a, b, c \in G$$

In this case, a and c are called 'conjugate' ($a \sim c$) or 'belonging to the same class'. Every set of conjugate group elements forms a class, the identity element always being a class by itself. In our example S_3 we have three classes (which are also shown in Table 7.1), namely $\{E\}$, $\{(12), (13), (23)\}$, and $\{(123), (132)\}$.

The following rules are valid for all classes: (i) classes never overlap; and (ii) the order of a class g is a divisor of the group order n.

7.2 Symmetry Groups

Symmetry groups are by far the most important groups for the chemist, both in quantum mechanics and also in the analysis of spectroscopic data. Therefore, the subsequent treatment of group theory and its methods will be focused on these groups and their representations.

7.2.1 Symmetry Operators

Symmetry groups contain symmetry operators as elements. These are

- $\hat{E}$, the identity operation, which does not change anything;
- $\hat{C}_n^k$, a proper axis, meaning the rotation about an axis by $360/n$ degrees;
- $\hat{S}_n^k$, an improper axis, meaning a rotation about an axis by $360/n$ degrees, followed by a reflection through a plane perpendicular to that axis;
- $\hat{\sigma}_v$, $\hat{\sigma}_h$, reflection through a vertical/horizontal plane; and
- $\hat{i}$, inversion through a center.

7.2.2 Symmetry Groups and their Representations

Symmetry groups consist of a number of symmetry operators as elements – the higher the symmetry, the larger the number of elements. The lowest symmetry thus consists only of two elements, namely $\hat{E}$ and $\hat{C}_1$, corresponding to a full rotation by 360 degrees. Symmetry groups in chemistry are usually characterised by the *Schönflies* symbols. Groups are classified as C_n groups if they contain a $\hat{C}_n$

element, as C_{nv} groups if they additionally contain a vertical symmetry plane, and as C_{nh} groups if there is also a horizontal symmetry plane. Additional symmetry elements lead to the groups D_n, D_{nh} and D_{nd}. There are, however, other groups outside these rules (e.g., the C_s and C_i groups) which, apart from $\hat{E}$, only have the $\hat{\sigma}$ or the $\hat{i}$ element, respectively. Groups with an improper rotation axis of a higher degree than any $\hat{C}_n$ element in the group are classified as S_n groups. Special Schönflies symbols are also used for the higher symmetries of the cubic groups, namely T, T_d and T_h (tetrahedral), O and O_h (octahedral), and for groups I and I_h (icosahedral). Particular cases of the C_{nv} and D_{nh} groups refer to linear molecules with a $\hat{C}_\infty$ rotation axis. The corresponding symmetry point groups are $C_{\infty v}$ and $D_{\infty h}$, where the latter additionally has an inversion center in the molecule (examples are the molecules CO and N_2).

If a symmetry operator $\hat{R}$ acts on a basis of vectors $\{e_i\}$, they will be transformed as outlined in Section 1.10. This transformation can be represented by a specific matrix according to $\hat{R}\mathbb{E} = \mathbb{E}\mathbb{R}$, where $\mathbb{R}$, the transformation matrix, is also called the representation of the symmetry operator $\hat{R}$ in the basis $\{e_i\}$. If we obtain the representation matrices for all symmetry operators of a group, they are together an equivalent representation of the group, and thus form a *group of representation matrices*. The relationship between the group of symmetry operators and the group of their representation matrices is in general a homomorphism, in some cases even an isomorphism, and we can exemplify the mappings from operators to matrices by

$$\hat{R} \to \mathbb{R}$$
$$\hat{1} \to \mathbb{1}$$
$$\hat{R}^{-1} \to \mathbb{R}^{-\mathbb{1}}$$
$$\hat{R}\hat{S} \to \mathbb{R}\mathbb{S}$$
$$\hat{R}\hat{R}^{-1} \to \mathbb{R}\mathbb{R}^{-\mathbb{1}} = \mathbb{1}$$

The group of the symmetry operators G_R is thus directly related to the group of matrices Γ_R, and the latter is called the *representation of the group*.

Symmetry operators are unitary operators (self-adjugate $R = R^*$), and thus their representation matrices are also unitary. They can be transformed by unitary transformations to another basis, and the basis can be extended, leading to extended representation matrices. If we have two bases $\{e_i\}$ and $\{f_i\}$, we can form the extended basis $\{g_i\}$, which contains the elements $(e_1, e_2, \dots e_n, f_1, f_2, \dots f_m)$, and the representation matrix $\mathbb{R}_g$ is given by the direct sum $\mathbb{R}_g = \mathbb{R}_e \oplus \mathbb{R}_f$.

Changing between bases of the same dimension by a unitary transformation does not change the trace of the representation matrix; that is, all equivalent representations have the same trace. The trace of representation matrices is, therefore, called the 'character' χ (this stability of character appears to be valid only in mathematical groups, not in societal 'groups').

The representation matrices of all symmetry operators belonging to the same class have the same character χ, as we can deduce from the definition of classes, where all elements are conjugated, i.e., they are related by similarity transformations.

7.2.3 Reducible and Irreducible Representations and Character Tables

Let us assume that we have an n-dimensional representation group Γ_R and the corresponding basis $\{g_i\}^n$ spanning the vector space V^n. This basis will be transformed by a similarity transformation with a matrix $\mathbb{U}$ to another basis $\{e_i\}^n = (e_1, e_2, \dots e_m, e_{m+1}, \dots e_n)$:

$$\mathbb{R}_e = \mathbb{U}^{-1}\mathbb{R}_g\mathbb{U}$$

Now, we assume that all symmetry operators will transform the vectors $(e_1 \cdots e_m)$ only within this subset, and the vectors $(e_{m+1} \cdots e_n)$ also only within that subset. Under this condition every representation matrix $\mathbb{R}_g$ will decompose into a direct sum of smaller matrices $\mathbb{R}' \oplus \mathbb{R}''$.

Thus, the representation group Γ_R has been reduced to smaller representations according to $\Gamma_R = \Gamma' \oplus \Gamma''$. Any representation that can be reduced this way is called a *reducible representation*. One can easily imagine that Γ' and/or Γ'' can be further reduced by similarity transformations, thus leading to a 'staircase'-formed direct sum of representations. When no further reduction to smaller representations is possible, the representations obtained are called *irreducible representations*. They will play a prominent role in all further group theoretical considerations. The general underlying scheme is

$$\Gamma_R = a_1\Gamma_1^i \oplus a_2\Gamma_2^i \oplus \cdots a_n\Gamma_n^i$$

Every irreducible representation can occur more than once, and the parameters a_n indicate how many times a specific irreducible representation is contained in the reducible one.

The Schönflies notation also provides specific symbols for the irreducible representations, which are commonly used in chemical literature and will, therefore, be summarised here:

A capital letter denotes the dimension:

- A,B........1-dimensional; A implies positive, B negative character for main rotational operation
- E..........2-dimensional
- T..........3-dimensional
- G..........4-dimensional
- H..........5-dimensional

- $'$.......positive character of $\hat{\sigma}_h$
- $''$......negative character of $\hat{\sigma}_h$

- $_1$.......positive character of a further element, e.g., $\hat{C}_2$ or $\hat{\sigma}_v$
- $_2$.......negative character of a further element, e.g., $\hat{C}_2$ or $\hat{\sigma}_v$

- g.......positive character of $\hat{i}$
- u.......negative character of $\hat{i}$

For the groups $C_{\infty v}$ and $D_{\infty h}$, special symbols are used, for which the following equivalences are valid:

$$A_1 \equiv \Sigma^+$$

$$A_2 \equiv \Sigma^-$$

$$E_1 \equiv \Pi$$

$$E_2 \equiv \Delta$$

$$E_3 \equiv \Phi$$

for the group $C_{\infty v}$, and in the group $D_{\infty h}$ these irreducible representations are split further, due to the inversion centre, to g and u types (e.g., Π_g and Π_u).

When physicists performed the first calculations leading to molecular orbitals for diatomic molecules, they classified these MOs according to the irreducible representations after which they transformed, using this nomenclature (i.e. as σ or π). When the chemists – who at that time were rather ignorant of group theory – saw these publications, they observed that molecules with single bonds had σ-MOs, and that there was an additional π-MO in double bonds. Thus, they associated single bonds with 'σ' electrons and double bonds with 'π' electrons and extended this nomenclature to all molecules, irrespective of the symmetry group to which they belonged. The so-called 'π' electrons of benzene are, therefore, simply an ill-used nomenclature – benzene is definitely not a linear molecule, but belongs to the symmetry group D_{6h} – and the 'π' – orbitals are correctly classified as A_{2u} and B_{2g} orbitals. This is another example of how a lack of theoretical knowledge can lead to an inadequate 'scientific' terminology.

Every symmetry group can be characterised by its elements or classes of elements, respectively, and its irreducible representations. These properties, together with the characters of the symmetry operations in the different irreducible representations, are summarised in so-called *character tables*:

Group symbol	**Class 1**	**Class 2**	**...**	**Class n**
IR_1	χ	χ	...	χ
IR_2	χ	χ	...	χ
...	χ	χ	...	χ
IR_n	χ	χ	...	χ

The character tables of some chemically important symmetry groups have been collected in Tables 7.2–7.4, which subsequently will be used to illustrate the usage of such groups.

Table 7.2 Character tables of the symmetry groups C_1, C_s, C_{2v}, C_{3v} and C_{4v}.

Character table of the group C_s.

C_1	E
A	1

Character table of the group C_S.

$C_s = C_h$	E	σ_h		
A'	1	1	x, y, R_z	x^2, y^2, z^2, xy
A''	1	-1	z, R_x, R_y	yz, xz

Character table of the group C_{2v}.

C_{2v}	E	C_2	$\sigma_v(xz)$	$\sigma_v'(yz)$		
A_1	1	1	1	1	z	x^2, y^2, z^2
A_2	1	1	-1	-1	R_z	xy
B_1	1	-1	1	-1	x, R_y	xz
B_2	1	-1	-1	1	y, R_x	yz

Character table of the group C_{3v}.

C_{3v}	E	$2C_3$	$3\sigma_v$		
A_1	1	1	1	z	x^2, y^2, z^2
A_2	1	1	-1	R_z	
E	2	-1	0	$(x, y)(R_x, R_y)$	$(x^2 - y^2, xy)(xz, yz)$

Character table of the group C_{4v}.

C_{4v}	E	$2C_4$	C_2	$2\sigma_v$	$2\sigma_d$		
A_1	1	1	1	1	1	z	$x^2 + y^2, z^2$
A_2	1	1	1	-1	-1	R_z	
B_1	1	-1	1	1	-1		$x^2 - y^2$
B_2	1	-1	1	-1	1		xy
E	2	0	-2	0	0	$(x, y)(R_x, R_y)$	(xz, yz)

Table 7.3 Character tables of the symmetry groups D_{3h}, $C_{\infty v}$ and T_d.

Character table of the group D_{3h}.

D_{3h} (6m2)	E	$2C_3$	$3C_2$	σ_h	$2S_3$	$3\sigma_v$		
A_1'	1	1	1	1	1	1		$x^2 + y^2, z^2$
A_2'	1	1	−1	1	1	−1	R_z	
E'	2	−1	0	2	−1	0	(x, y)	$(x^2 - y^2, xy)$
A_1''	1	1	1	−1	−1	−1		
A_2''	1	1	−1	−1	−1	1	z	
E''	2	−1	0	−2	1	0	(R_x, R_y)	(xz, yz)

Character table of the group $C_{\infty v}$.

$C_{\infty v}$	E	$2C_\infty^\phi$	$\cdots$	$\infty\sigma_v$		
$A_1 \equiv \Sigma^+$	1	1	$\cdots$	1	z	$x^2 + y^2, z^2$
$A_2 \equiv \Sigma^-$	1	1	$\cdots$	−1	R_z	
$E_1 \equiv \Pi$	2	$2\cos\phi$	$\cdots$	0	$(x, y)(R_x, R_y)$	(xz, yz)
$E_2 \equiv \Delta$	2	$2\cos 2\phi$	$\cdots$	0		$(x^2 - y^2, xy)$
$E_3 \equiv \Phi$	2	$2\cos 3\phi$	$\cdots$	0		
$\cdots$	$\cdots$	$\cdots$	$\cdots$			

Character table of the group T_d.

T_d (43m)	E	$8C_3$	$3C_2$	$6S_4$	$6\sigma_d$		
A_1	1	1	1	1	1		$x^2 + y^2 + z^2$
A_2	1	1	1	−1	−1		
E	2	−1	2	0	0		$(2z^2 - x^2 - y^2, x^2 - y^2)$
T_1	3	0	−1	1	−1	$(R_x, R_y R_z)$	
T_2	3	0	−1	−1	1	(x, y, z)	(xy, xz, yz)

The *properties of irreducible representations* will be of importance for all group theoretical applications presented here, and hence will be summarised in the following:

1. Dimensions and group order: The dimensions l_i of all irreducible representations are related to the order of the group by

$$\sum_i l_i^2 = g$$

Table 7.4 Character table of the symmetry group O_h.

O_h ($m3m$)	E	$8C_3$	$6C_2$	$6C_4$	$3C_2$ ($=C_4^2$)	i	$6S_4$	$8S_8$	$3\sigma_h$	$6\sigma_d$		
A_{1g}	1	1	1	1	1	1	1	1	1	1		$x^2+y^2+z^2$
A_{2g}	1	1	−1	−1	1	1	−1	1	1	−1		
E_g	2	−1	0	0	2	2	0	−1	2	0		$(2z^2-x^2-y^2, x^2-y^2)$
T_{1g}	3	0	−1	1	−1	3	1	0	−1	−1	(R_x, R_y, R_z)	
T_{2g}	3	0	1	−1	−1	3	−1	0	−1	1		(xz, yz, xy)
A_{1u}	1	1	1	1	1	−1	−1	−1	−1	−1		
A_{2u}	1	1	−1	−1	1	−1	1	−1	−1	1		
E_u	2	−1	0	0	2	−2	0	1	−2	0		
T_{1u}	3	0	−1	1	−1	−3	−1	0	1	1	(x, y, z)	
T_{2u}	3	0	1	−1	−1	−3	1	0	1	−1		

2. Representation matrices: The irreducible representations find a 1:1 correspondence in representation matrices

$$\Gamma_i, \Gamma_j \leftrightarrow \mathbb{R}^i, \mathbb{R}^j$$

For the elements of these representation matrices the following relation is valid, which means that columns from different irreducible representations are orthogonal:

$$\sum_R R_{rs}^{i*} R_{tu}^j = \frac{g}{\sqrt{l_i l_j}} \delta_{ij} \delta_{rt} \delta_{su}$$

3. Characters: We can now study the product of characters from different irreducible representations

$$\chi^i = \sum_{r=1}^{l_i} R_{rr}^i, \ \chi^j = \sum_{s=1}^{l_j} R_{ss}^j$$

By forming the sum over all symmetry elements we come to the following relationships:

$$\sum_r \chi_{(r)}^{i*} \chi^i = \sum_{r=1}^{l_i} \sum_{s=1}^{l_j} \sum_R R_{rr}^{i*} R_{ss}^j = \sum_{r=1}^{l_i} \sum_{s=1}^{l_j} \frac{g}{\sqrt{l_i l_j}} \delta_{ij} \delta_{rs} = \sum_{r=1}^{l_i} \frac{g}{l_i} \delta_{ij} = g\delta_{ij}$$

We can sum over classes instead of elements as all elements in a class have the same character, h_c denoting the number of elements in the class, and this yields

$$\sum_c h_c \chi_{(c)}^{i*} \chi^j = g\delta_{ij}$$

the orthogonality relation of the characters, and the relationship between characters in classes and the group order

$$\sum h_c \chi_c^{(i)*} \chi^i = g$$

4. Classes and irreducible representations: The number of non-equivalent irreducible representations of a group is always equal to the number of classes in that group. A special irreducible representation in each group is the one where the character of all classes equals 1. This is called the *totally symmetric irreducible representation*.

Based on these properties one can develop a simple practical procedure for the *reduction of reducible representations*, based on the characters χ_i instead of performing a number of subsequent similarity transformations $\mathbb{U}^{-1}\mathbb{R}\mathbb{U}$ (remember that similarity transformations do not change the trace of a matrix, and thus not the character χ):

$$\chi_{\hat{R}} = \sum_j a_j \chi^j_{(R)}$$

where $\chi_{\hat{R}}$ is the character of $\hat{R}$ in the reducible representation Γ_{red}; a_j is the number, how many times the irreducible representation is contained in Γ_{red}; and $\chi^j_{(R)}$ is the character of the element $\hat{R}$ in the jth irreducible representation.

We can now use the orthogonality relation of the characters

$$\sum_{R \in G} \chi^{i*}_{(R)} \chi^j_{(R)} = \sum_j a_j \sum_{R \in G} \chi^{i*}_{(R)} \chi^j_{(R)}$$

to derive the expression which tells, how many times an irreducible representation occurs in a reducible representation:

$$a_i = \frac{1}{g} \sum_{\hat{R}} \chi^{i*}_{(R)} \chi_{(R)} = \frac{1}{g} \sum_C h_c \chi^{i*}_{(C)} \chi_{(C)}.$$

We will demonstrate the use of this expression with an example of the group C_{3v}, the character table of which is given in Table 7.2. Let us assume we have a reducible four-dimensional representation Γ_r with the characters 4,1 and 0 for the classes E, C_3 and σ (the character of E always indicates the dimension of the representation). The evaluation of the a_i values delivers:

$$a_{A_1} = \frac{1}{6}(1 \cdot 1 \cdot 4 + 2 \cdot 1 \cdot 1 + 3 \cdot 0 \cdot 1) = 1$$

$$a_{A_2} = \frac{1}{6}(1 \cdot 1 \cdot 4 + 2 \cdot 1 \cdot 1 + 3 \cdot 0 \cdot -1) = 1$$

$$a_E = \frac{1}{6}(1 \cdot 2 \cdot 4 + 2 \cdot 1 \cdot -1 + 3 \cdot 0 \cdot 0) = 1$$

and hence we have obtained the composition of Γ_r as $\Gamma_r = A_1 \oplus A_2 \oplus E$.

Under certain conditions, two groups can be combined by a direct multiplication, and the result is the *direct product of groups*. The conditions to enable such a direct product are either the identity of both groups, $G_1 = G_2$, or the relationship

$$G_{1i}G_{2j} = G_{1j}G_{2i} \quad \forall i, j$$

In the latter case, the two groups are called 'independent' groups.

Corresponding to the direct product $G_P = G_1 \otimes G_2$ the representations and all representation matrices are obtained as $\Gamma_P = \Gamma_1 \otimes \Gamma_2$ and $\mathbb{R}_{iP} = \mathbb{R}_{i1} \otimes \mathbb{R}_{i2}$. The trace of the direct product of two matrices $\mathbb{A} \otimes \mathbb{B}$ is given by $tr\ \mathbb{A} \cdot tr\ \mathbb{B}$, and thus all characters in the product group can be calculated as the product of the characters in the respective irreducible representations.

Within a group it is also possible to construct reducible representations by direct multiplication of irreducible representations. For example, the direct product of the irreducible representation E in the group C_{3v} with itself results in the reducible representation Γ_r with the characters 4,1 and $\emptyset$, the reduction of which was shown in the previous example. We will encounter such direct products of irreducible representations again in the applications of group theory.

As an example for the direct product of two independent groups we will consider the groups C_{3v} and C_s. The direct product of the group elements means the multiplication of each element of the first group with each element of the second one. In our example, the elements $C_{3v} : [E\ 2C_3\ 3\sigma_v]$ and the elements $C_S : [E\ \sigma_h]$ give

$$C_{3v} \otimes C_s = [E\ 2C_3\ 3\sigma_v\ E \cdot \sigma_h\ \overbrace{2C_3\sigma_h}^{2S_3}\ \overbrace{3\sigma_v\sigma_h}^{3C_2}],$$

and we can compare this result with the character tables (Table 7.3), which tells us that we have obtained the group D_{3h}. Actually, we can easily construct the full character table of this group by the direct multiplication of irreducible representations of the groups C_{3v} and C_s. By recalling the character tables of these groups (Table 7.2):

C_{3v}	E'	$2C_3$	$3\sigma_v$
A_1	1	1	1
A_2	1	1	−1
E	2	−1	0

C_s	E	σ_h
A'	1	1
A''	1	−1

We create the individual irreproducible representations, and the corresponding characters as shown below:

D_{3h}	E $= E \cdot E'$	$2C_3$ $= E \cdot C_3$	$3\sigma_v$ $= E \cdot \sigma_v$	σ_n $= E' \cdot \sigma_h$	$2S_3$ $= 2C_3 \cdot \sigma_h$	$3C_2$ $= 3\sigma_v \cdot \sigma_h$
$A'_1 = A_1 \otimes A'$	1	1	1	1	1	1
$A''_1 = A_1 \otimes A''$	1	1	1	−1	−1	−1
$A'_2 = A_2 \otimes A'$	1	1	−1	1	−1	−1
$A''_2 = A_2 \otimes A''$	1	1	−1	−1	−1	1
$E' = E \otimes A'$	2	−1	0	2	−1	0
$E'' = E \otimes A''$	2	−1	0	−2	1	0

A comparison with the original character table of group D_{3h} in Table 7.3 confirms that we have obtained the correct description of this group. By the direct product of two groups we have obtained a new group of higher dimension and thus performed an ascent in symmetry. In chemistry we will often encounter the opposite effect – namely a descent in symmetry – when by configurational changes or substitutions a molecule of high symmetry descends to lower symmetry; for example, an octahedral compound by elongation of two opposite bonds to a tetragonal bipyramid of D_{4h} symmetry. Such changes of symmetry are always accompanied by changes in the MO energies and the associated spin orbitals and changes of the vibrational spectra of the molecules. We will see some of these effects in the following examples for group theory applications in chemistry.

7.3 Applications of Group Theory in Quantum Chemistry

Group theoretical considerations can predict whether any integral of the types

$$\langle \Phi | \hat{O} \Phi \rangle \quad \text{or} \quad \langle \Phi_1 | \hat{O} \Phi_2 \rangle$$

i.e., expectation values or transition elements can produce a value $\neq \emptyset$. For this purpose it is only necessary to know the irreproducible representations, to which the functions Φ and the operator $\hat{O}$ belong, and to perform a direct multiplication of these irreproducible representations, e.g. for the case $\langle \Phi_1 | \hat{O} \Phi_2 \rangle$ the direct product $\Gamma_{\Phi_2} \otimes \Gamma_O = \Gamma_P$.

If the product representation Γ_P does not contain Γ_{Φ_1}, the integral $\langle \Phi_1 | \hat{O} \Phi_2 \rangle$ is zero. This is a group theoretical formulation of the '*no combination rule*', which is also expressed by the condition that the reduction of Γ_I, defined by:

$$\Gamma_I = \Gamma_{\Phi_1} \otimes \Gamma_O \otimes \Gamma_{\Phi_2}$$

must contain the totally symmetric irreproducible representation Γ_{tot} to yield a value $\neq \emptyset$ for the integral. Γ_{tot} is always the first row in a character table, and all of its χ-values are 1.

If the operator's irreproducible representation is Γ_{tot} – and this is the case for the Hamiltonian! – the judgement of vanishing integrals is even easier. As the direct product of any irreproducible representation with Γ_{tot} yields the same irreproducible representation,

$$\Gamma_i \otimes \Gamma_{tot} = \Gamma_i$$

Φ_1 and Φ_2 must belong to the same irreproducible representation to yield a non-zero integral $\langle \Phi_1 | \hat{O}\Phi_2 \rangle$. Thus, symmetry considerations can remarkably reduce the computational effort needed in the evaluation of integrals in a quantum mechanical calculation.

The use of symmetry-adapted functions can also facilitate such calculations, and it is possible to construct such functions with the help of 'symmetry projectors'

$$\hat{P}^j = \frac{l_j}{g} \sum_{\hat{R}} \chi_{\hat{R}}^j \hat{R}$$

where j denotes the jth irreproducible representation of the group, $\chi_{\hat{R}}^j$ the character of the symmetry operation $\hat{R}$ in this representation, and the summation runs over all symmetry operations of the group. This can be illustrated by a small example for the group C_i.

There are two symmetry elements in the group, $\hat{E}$ and $\hat{i}$. Acting on a function $f(x)$ they produce

$$\hat{E}f(x) = f(x) \quad \text{and} \quad \hat{i}f(x) = f(-x)$$

The group C_i provides two irreducible representations, A_g and A_u with the characters $(1, 1)$ and $(1, -1)$, and we will make use of the appropriate projectors to produce functions transforming according to these representations:

$$\hat{P}^{A_g} = \frac{1}{2}[1 \cdot f(x) + 1 \cdot f(-x)] = \phi_g$$

$$\hat{P}^{A_u} = \frac{1}{2}[1 \cdot f(x) - 1 \cdot f(-x)] = \phi_u$$

By applying the inversion operator $\hat{i}$ to ϕ_u one can easily prove that this function indeed transforms according to A_u:

$$\hat{i}\phi_u = \hat{i}\frac{1}{2}[f(x) - f(-x)] = \frac{1}{2}[f(-x) - f(x)] = -\phi_u$$

7.4 Applications of Group Theory in Spectroscopy

There are manifold ways to make use of group theory in spectroscopy, based on the consideration of transition elements $\langle \Phi_1 | \hat{O}\Phi_2 \rangle$, which describe the transition from a state $|\Phi_1\rangle$ to the state $|\Phi_2\rangle$. The operator $\hat{O}$ specifies the type of transition; for example, in electronic transitions it would be the operator of the dipole transition $\hat{\mu} = (xyz)$.

Even without considering specific quantum mechanically described states, group theory is extremely useful to provide a qualitative prediction of vibrational spectra. In order to illustrate these capabilities, two examples will be treated in the following sections.

7.4.1
Example 1: Electron Spectroscopy

Here, we will present a qualitative treatment of the so-called $d \rightarrow d$ transitions of a metal ion in a tetrahedral environment, corresponding to the ligand field model discussed earlier. According to this model, the five levels of the d-electrons are split into a lower 2-fold degenerate E-level and a higher three-fold degenerate T_2 level, as illustrated in Fig. 7.1 (the nomenclature of these levels corresponds to the irreducible representations in the tetrahedral symmetry group).

Let us assume now the simple case of an ion with two d-electrons (e.g. V^{3+}), where both electrons are in the lower E-level, and where one of them will be excited to a T_2 level. First, we must determine the possible irreducible representations for the ground state e^2 and the excited state $e^1{t_2}^1$:

$$\Gamma_{e^2} = E \otimes E = A_1 \oplus A_2 \oplus E$$

$$\Gamma_{e^1{t_2}^1} = E \otimes T_2 = T_1 \oplus T_2.$$

Due to the Pauli principle, only the electronic states ${}^1A_1$, ${}^3A_2$, ${}^1E$ are allowed for the ground state, as illustrated in Fig. 7.1, whereas for the excited state all possible spin states are allowed, namely ${}^1T_1$, ${}^3T_1$, ${}^1T_2$, ${}^3T_2$.

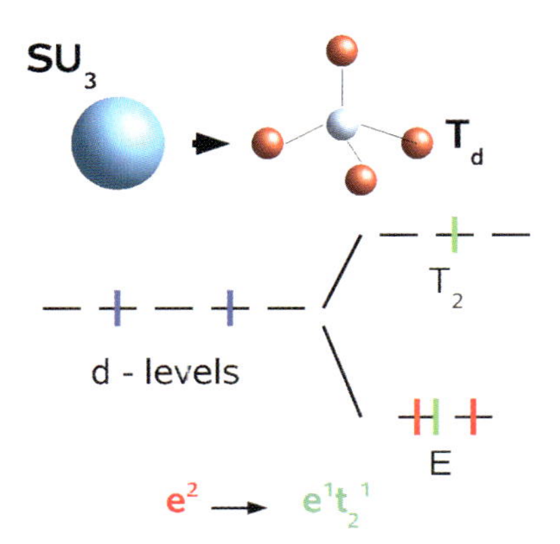

Fig. 7.1 Splitting of d-functions in tetrahedral ligand field for d^2 configuration.

The operator of the transition $\hat{\mu}$ transforms according to the irreducible representation T_2. Thus, we must form the product of this irreducible representation with those of the ground state configurations:

$$A_1 \otimes T_2 = T_2 \quad A_2 \otimes T_2 = T_1 \quad E \otimes T_2 = T_1 \oplus T_2.$$

We see that both irreducible representations of the excited state are present in these products, and can thus predict four allowed electronic transitions, by applying the spin selection rule for such transitions, $\Delta S = 0$, i.e., conservation of the total spin S:

$$e^2\,{}^1A_1 \rightarrow e^1 t_2{}^1\,{}^1T_2$$
$$e^2\,{}^3A_2 \rightarrow e^1 t_2{}^1\,{}^3T_1$$
$$e^2\,{}^1E \rightarrow e^1 t_2{}^1\,{}^1T_1$$
$$e^2\,{}^1E \rightarrow e^1 t_2{}^1\,{}^1T_2$$

7.4.2
Example 2: Infrared/Raman Spectroscopy

Group theory can be used as a tool to determine how many IR- and Raman-active molecular vibrations a molecule has, and to assign these vibrations – and thus the observed spectral bands – to irreducible representations. On the other hand, if the structure of the molecule is not known, but only its composition, the prediction of IR/Raman lines for different possible structures can help to determine the structure of the molecule. We will illustrate this with two simple examples, BF_3 and N_2F_2 (Fig. 7.2).

In the case of BF_3, the first step is to find a reducible representation of the molecule, which is achieved through the Cartesian coordininates of the atoms, and the changes of these coordinates under the symmetry operation classes of the point group to which the molecule belongs (in this procedure, atoms changing their positions upon a symmetry operation do not have to be considered!). The effect of the symmetry operations on the molecule BF_3, which belongs to the point group D_{3h}, is illustrated in Fig. 7.3.

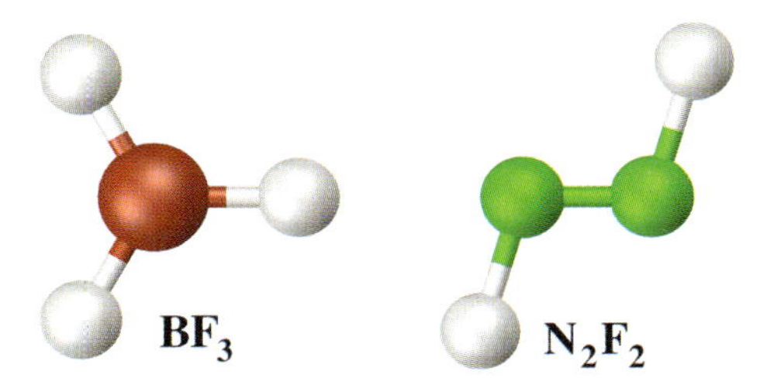

Fig. 7.2 Structure of molecules BF_3 and N_2F_2.

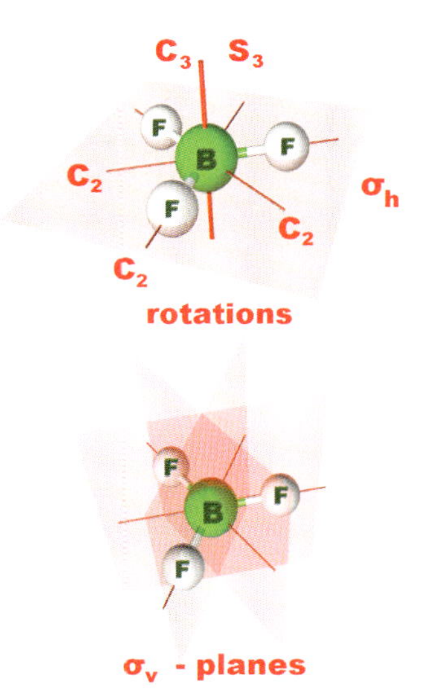

Fig. 7.3 Symmetry elements of the group D_{3h} in the example BF_3.

In the case of $\hat{E}$, all atoms remain at their original location, and as we have 12 Cartesian coordinates x, y, z for the four atoms, $\hat{E}$ is represented by a unit matrix of the dimension 12. Thus, the character χ for $\hat{E}$ in the new representation is 12.

When a $\hat{C}_3$ operation is performed, all fluorine atoms change their position, and we thus must consider only the coordinates of the central boron atom (one has to consider the atom as an item of finite dimension, not just a point!). For them, the rotation by 120 degrees is represented by the rotation matrix

$$\begin{pmatrix} -\frac{1}{2} & -\frac{\sqrt{3}}{2} & 0 \\ \frac{\sqrt{3}}{2} & -\frac{1}{2} & 0 \\ 0 & 0 & 1 \end{pmatrix}$$

and the character χ, therefore, results as $\emptyset$.

A $\hat{C}_2$ operation changes the position of two fluorine atoms, and we must consider the coordinates of the third fluorine and the boron atom. For both a rotation matrix with trace -1 results in

$$\begin{pmatrix} -1 & 0 & 0 \\ 0 & 1 & 0 \\ 0 & 0 & -1 \end{pmatrix}$$

and the character of both together therefore results as $\chi = -2$.

When $\hat{\sigma}_h$ is performed, all x and y coordinates remain unchanged, and all z coordinates change their sign, corresponding to four matrices

$$\begin{pmatrix} 1 & 0 & 0 \\ 0 & 1 & 0 \\ 0 & 0 & -1 \end{pmatrix}$$

and a total character $\chi = 4$.

The improper rotation $\hat{S}_3$ changes all fluorine coordinates. For the boron coordinates we can construct the matrix representing the changes from the product of the matrix for $\hat{C}_3$ and that for $\hat{\sigma}_h$, resulting in

$$\begin{pmatrix} -\frac{1}{2} & 0 & 0 \\ 0 & -\frac{1}{2} & 0 \\ 0 & 0 & -1 \end{pmatrix}$$

and a total character $\chi = -2$.

Finally, $\hat{\sigma}_v$ leaves one fluorine atom and boron in their places and their coordinate transformations are characterised by the matrices

$$\begin{pmatrix} 1 & 0 & 0 \\ 0 & -1 & 0 \\ 0 & 0 & 1 \end{pmatrix}$$

leading to a total value $\chi = 2$.

Thus, we can summarise the reducible representation $\hat{\Gamma}$ by the following character values:

$\hat{E}$	$\hat{C}_3$	$\hat{C}_2$	$\hat{\sigma}_h$	$\hat{S}_3$	$\hat{\sigma}_v$
12	0	−2	4	−2	2

The reduction according to the formula given in Section 7.2.3 leads to the following irreducible representations:

$$\Gamma = A_1' \oplus A_2' \oplus 3E' \oplus 2A_2'' \oplus E''$$

We must now consider the translations and rotations up to the maximum of the degrees of freedom (3n − 6 for a non-linear molecule with n atoms). The last column in all character tables lists, according to which irreducible representation the vectors x, y, z, the rotators R_x, R_y, R_z and the tensors composed of x, y, z vectors, e.g. xy or $x^2 + y^2$ transform. On examining the character table of the group D_{3h} in Table 7.3 we see that we must subtract from the irreducible representations of $\hat{\Gamma}$ A_2' for R_z and E'' for R_x, R_y. For the translations we must take out one of the three E' and one of the two A_2''. The remaining irreducible representations are available for molecular vibrations:

$$\Gamma = A_1' \oplus 2E' \oplus A_2''$$

This means that we should observe four bands in the vibrational spectrum of BF_3. The character table further helps us to identify these vibrations as IR- or Raman-active. IR-active vibrations transform as the vectors, and Raman-active vibrations as the tensors in the tables. Thus, the vibration according to A_1' and the two vibrations according to E' are seen in the Raman spectrum, whereas the vibration according to A_2'' will be visible in the IR spectrum.

Our second example is the molecule N_2F_2, which belongs to symmetry group C_{2h} (cf. Fig. 7.2) with the character table

C_{2h}	E	C_2	i	σ_h		
A_g	1	1	1	1	R_z	x^2, y^2, z^2, xy
B_g	1	−1	1	−1	R_x, R_y	xz, yz
A_u	1	1	−1	−1	z	
B_u	1	−1	−1	1	x, y	

The different symmetry operations yield the following character values:

- $\hat{E}$: all coordinates unchanged $\Rightarrow \chi = 12$
- $\hat{C}_2$: all atoms moved $\Rightarrow \chi = 0$
- $\hat{i}$: all atoms moved $\Rightarrow \chi = 0$
- $\hat{\sigma}_h$: all x and y coordinates unchanged, all z $\rightarrow$ −z, leading to four matrices with the trace $(1\ 1\ -1) \Rightarrow \chi = 4$

Thus, the reducible representation results as

$\hat{E}$	$\hat{C}_2$	$\hat{i}$	$\hat{\sigma}_h$
12	0	0	4

which, after reductions, yields

$$\Gamma = 4A_g \oplus 2B_g \oplus 2A_u \oplus 4B_u$$

For rotations, we must subtract one A_g and two B_g, for translation one A_u and two B_u. The remaining three A_g lines are Raman-active; the one A_u and the two B_u vibrations present can only be observed in the IR spectrum.

Although the example applications discussed here have been chosen to demonstrate the versatility of group theory in chemistry, it should have become clear by now that modern theory in chemistry cannot be carried out without group theory. The notation of *photoelectron spectra (PES)* underlines this statement: Following Koopmans' theorem (*vide supra*), the electronic energies are associated with the one-particle energies of the canonical MOs, and hence the spectral lines are assigned to these MOs. The notation of these MOs follows their irreducible representations, and one must determine, therefore, for each of the canonical MOs the irreducible representation after which it transforms.

Test Questions Related to this Chapter

1. In what types of molecule can we have σ- and π-electrons?
2. What is the irreducible representation of the Hamiltonian in the groups O_h, T_d and C_s?
3. How can we decide, whether a molecular vibration is IR- or Raman-active?
4. How can we decide, on a group theoretical basis, whether the integral $\langle \Phi_1 | \hat{H} \Phi_2 \rangle$ has a non-zero value?
5. Why are the permutation groups S_n exemplary for all other finite groups?
6. Why is it mathematically wrong – according to the group postulates – to talk about 'group sex'?

8
Computational Quantum Chemistry Methods

From nothing we can better recognise the lack of mathematical knowledge than from an exaggerated accuracy in numerical calculations.

Carl Friedrich Gauss, 18th century

This chapter will provide an overview of the most common methods used in computational chemistry. In following the intention of this book to provide a basic introduction into theoretical and computational chemistry, this overview can supply only a rather general and even partly superficial insight into the methods, sufficient to know the principles and to understand for which purposes they can be utilised (and for which they should *not* be applied!). For an in-depth knowledge of the methods and all of their variations, the reader should consider one or more of the specialised books dealing with computational chemistry procedures and programmes.

8.1
Ab Initio Methods

The terms '*ab initio*' and '*first principle*' are not always used in a standardised way. We will, therefore, stick to a rather strict and unambiguous nomenclature, reserving these terms for methods in which no fitting parameters of any kind nor any empirical approximations whose impact cannot be controlled by methodically-inherent procedures, are used. All methods *not* fulfilling these conditions are, therefore, described in the other sections of this chapter.

8.1.1
Ab Initio Hartree–Fock (HF) Methods

The foundations of this method have already been provided in detail in Chapter 5, while presenting the Hartree–Fock formalism in the LCAO-MO approach, leading to the Roothaan–Hall equation: $\mathbb{F}\mathbb{C} = \mathbb{S}\mathbb{C}\mathbb{E}$.

The Basics of Theoretical and Computational Chemistry. Edited by B. M. Rode, T. S. Hofer and M. D. Kugler

ISBN: 978-3-527-31773-8

It has also been mentioned that, in order to solve the equation, one must perform a Löwdin orthonormalisation procedure first to convert the $\mathbb{S}$-matrix to the unit matrix, then to build the $\mathbb{F}$-matrix and to diagonalise it to obtain the eigenvalue matrix $\mathbb{E}$ and the eigenvector matrix $\mathbb{C}$.

In order to build the Fock matrix with its elements:

$$F_{rs} = H_{rs}^{c} + \sum_{t}\sum_{u} P_{tu}\left[(rs/tu) - \tfrac{1}{2}(rt/su)\right],$$

one needs to know a set of initial linear combination coefficients for the calculation of the density matrix elements P_{tu}, and we have seen that these coefficients must be improved in the SCF process until consistency is achieved:

$$\mathbb{C}_0 \Rightarrow \mathbb{P}_0 \Rightarrow \mathbb{F}_0 \Rightarrow \mathbb{C}_1 \Rightarrow \mathbb{P}_1 \Rightarrow \mathbb{F}_1 \Rightarrow\Rightarrow\Rightarrow \mathbb{C}_n \Rightarrow \mathbb{P}_n \Rightarrow \mathbb{F}_n$$

The practical process of HF *ab initio* calculations thus consists of the following steps:

1. Choice of a basis set for all atoms in the system to be calculated, and selection of the calculation method and spin state of the system.
2. Calculation of the overlap integral matrix $\mathbb{S}$.
3. Performance of the Löwdin orthonormalisation with $\mathbb{S}^{-\frac{1}{2}}$.
4. Evaluation of the elements of the core Hamiltonian H_{rs}^{c}.
5. Diagonalisation of the $\mathbb{H}^{c}$ matrix, leading to a starting set for the coefficients $\mathbb{C}_{\circ}$ and an initial density matrix $\mathbb{P}_{\circ}$ (other methods for this initial guess are also available).
6. Evaluation of all two-electron integrals (rs/tu), (rt/su).
7. Construction of the initial Fock matrix $\mathbb{F}_{\circ}$.
8. Iterative SCF procedure.
9. Evaluation of results.

From step (2) onwards, almost everything is managed by modern quantum chemical program packages more or less automatically, and simply controlled by a few parameters or keywords that the user has to provide. This explains why it has become so easy to perform such calculations without a sound theoretical background – and then to give an incompetent interpretation to the results!

The first of these steps – the choice of the basis set – is the most critical one, as it determines the quality and significance of the results. On the other hand, it has a strong influence on the computational effort needed, as this effort increases approximately with n^4, if n is the number of basis functions. The literature is replete with basis sets for all types of atoms, and commercial *ab initio* programmes contain libraries of large numbers of standard basis sets; hence, the decision of which basis set to employ is left to the user. It seems important, therefore, to deal with basis sets in more detail here, as this choice is – besides the ignorance of the calculation method itself – the most frequent error source for users of 'black box' quantum chemical programs. This is true not only for HF *ab initio* calculations, but also for correlated post-HF methods and density functional methods.

The types of basis sets and some of the problems associated with them have already been mentioned in Chapter 5, where it was also stated that contemporary quantum mechanical calculations mostly employ Gauss-type orbitals (GTOs):

$$\varphi_{GTO} = N \cdot e^{-\alpha \cdot r^2} \cdot Y_{\theta,\varphi}^{l,m}$$

as 'primitive' basis functions, which are combined to represent the traditional Slater functions in a numerically more favourable form. Therefore, one of the classifications of basis sets indicates how many GTOs are used to represent one STO for example, by STO-3G or STO-6G (the latter was used in the formamide example and is listed in Table 5.1, although without the fixed coefficients by which the six functions used to represent the corresponding STO are combined). Such basis sets, where one GTO-constructed STO is used for each valence shell corresponding to the electronic scheme of the periodic system, are called *single-zeta basis sets*. If each of these shells is represented by two or three STOs, one calls the basis sets *double- or triple-zeta basis sets*, respectively. Examples are the 6-31G and 6-311G basis sets, where the inner electron-STOs are represented by six GTOs. For the valence electrons, two or three independent basis functions are used, one of them composed of three GTOs, while the other one (two) are independent GTOs. It is clear that double- and triple-zeta basis sets ensure an increasing flexibility in the construction of the spin orbitals, and thus a better description of the chemical system – albeit at the cost of greatly increased computer time.

In many cases, double- or triple-zeta basis sets are insufficient for an adequate description of the physical and chemical properties of molecules, especially when polarisation effects play an essential role in these properties. In such cases the use of additional '*polarisation functions*' is mandatory. These are higher functions, perhaps p-functions for hydrogen and d-functions for second-row elements such as carbon, nitrogen, and oxygen. Sometimes, even the use of f- and g-functions is needed.

On the other hand, studies of excited states or of anions can require the use of so-called '*diffuse functions*', allowing a better description of electron distribution far from the nucleus.

In addition to a more flexible and appropriate description of the system extension of the basis set generally also leads to an improved (i.e., lower) electronic and total energy value. This improvement with the number of basis functions becomes gradually smaller, converging asymptotically to a value where the further addition of functions has no more effect on the energy. This limit is called the '*Hartree–Fock limit*', and only correlated methods (which will be discussed in the next section) can overcome this limit.

In most practical cases one operates quite far from the HF limit, and other considerations in the choice of the basis set become stringent. First, one must take care that all atoms are described in a balanced way; that is, the number of basis functions assigned to the atoms corresponds (roughly) to the number of electrons they contribute to the system. The many standard basis sets contained in the libraries of available commercial programmes take care of this, but when individual basis sets are assigned or constructed, one must consider this condition of *balanced basis sets.*

Another important phenomenon to be considered is the *basis set superposition error (BSSE)*, which can result from the use of smaller basis sets. This error becomes relevant, when the interaction of two molecules is investigated by calculating the isolated molecules first, and then the adduct of them. In the latter case, the basis functions of one partner can act as an 'improvement' of the basis of the other, thus leading to a mutual artificial improvement of the description and an artificial stabilisation. An estimation of the upper limit for this error can be made using the *Boys–Bernardi* procedure, where each of the partners is calculated with the (empty) functions of the other, located at the positions that they assume in the adduct. The additional stabilisations achieved in these calculations together indicate how large the BSSE effect could become.

The heavier the atoms become, the more functions must be assigned to them, and the larger becomes the computational effort. On the other hand, we know that most inner electrons in heavy atoms do not contribute much to the chemistry of these atoms in a molecule (i.e., to binding and reactivity). A very convenient and common way to make use of this fact is to describe the inner electrons by a few specific functions that represent them, the so-called *effective core potentials (ECP)*; these are sometimes also referred to (in a less appropriate way) as *pseudopotentials* (the Greek word ψεύδω means 'to lie'!). In practice, the ECP will describe the next lower rare gas configurations; for example, in the case of first row transition elements the [*Ar*] core, or in the case of Ag or I the [*Kr*] core. Together with the ECPs, adapted valence-shell basis sets have been developed, and the combination of ECP and double-zeta valence basis sets leads to equivalent or even better results than all-electron basis set calculations. The reason for these better results is the possibility of taking into account *relativistic effects* in the construction of the ECPs. We already know that these relativistic contributions have an influence on the behaviour of valence electrons (e.g., bond shortening), and it is thus advantageous to include them into the calculations of systems with heavier atoms. Thus, the use of ECPs not only significantly reduces the computational effort, but also makes all atoms in the periodic system accessible to adequate *ab initio* calculations.

Another important aspect of quantum chemical calculations is the geometry of the molecule(s) in the calculated system. In some cases, the experimental geometry can be used as input, but if such data are not available, or if a comparison of the method-inherent geometry prediction with experimental data is desired, one must perform a *geometry optimisation* within the *ab initio* framework, which means evaluating a multidimensional energy surface. Due to the Born–Oppenheimer approximation, this surface must be calculated pointwise, though modern quantum mechanical programmes provide automated procedures for this purpose, mostly based on a scan of the energy surface and subsequent gradient calculations. According to the forces calculated to act on each atom, the positions of the atoms are changed in the opposite direction of the gradient, and this is repeated until all atoms are in the equilibrium positions, where the total energy of the molecule(s) is a minimum. As this is quite a computer-time intensive procedure (a gradient calculation takes about three times longer than the energy calculation), such optimisations are often performed with smaller basis sets, and a large basis set is only employed to calculate the optimised structure. Second-order gradient calculations supply vibrational frequencies for stretching and deformation of bond lengths and angles, respectively.

Having finished the calculation, a large amount of data will result. Besides the total energy, one-particle energies, canonical and/or localised molecular orbitals, atomic and overlap populations can be printed out, dipole and multipole moments, and many others. The evaluation and assessment of the validity of these data requires the knowledge of all the methodical framework and its error sources, and is by no means a trivial task. When associating the results with the chemical behaviour of compounds under laboratory conditions or in nature, one should never forget that the calculations have been made for the gas phase and at zero temperature. Some calculation methods allow ways to correct this, for example by adding an empirical term for a solvent influence, and often the analogy between gas-phase data and 'reality' is quite good. However, comparisons with calculations in solution and at room temperature can reveal enormous differences in many cases; examples will be discussed in the section of statistical simulations.

In order to illustrate *ab initio* HF calculations, an example is given in Appendix 1, showing a calculation of hyposulfuric acid H_2SO_3, including optimisation of the geometry. The data displayed show the input of starting geometry and a DZP basis set, the result of the geometry optimisation by the gradient method (cf. Fig. 8.1) the SCF cycles with the optimised geometry, and the final geometry and energy of the molecule. Subsequently, the one-particle eigenvalues (occupied and virtual) are shown, and finally the results of the Mulliken population analysis and dipole/quadrupole moments.

A final important feature of *ab initio* HF calculations is the handling of other than closed-shell electron configurations. It is not only molecules with an *uneven* number of electrons (e.g., NO) in which the ground state is an *open-shell configuration* with unpaired electrons, but also for molecules with an *even* number of electrons an open-shell configuration can be the most stable. In addition to the most famous example of O_2, many metal complexes in high-spin configuration prove this point. Naturally, most excited states of molecules are also open-shell configurations. There are two principal methods to deal with such open-shell systems within the HF scheme.

The first approach treats the paired electrons as a closed-shell subsystem, which is combined with additional spatial spin orbitals with single occupation. Figure 8.2 illustrates this in a graphical manner, showing possible occupations of the spin orbitals symbolised by one-particle energy levels. This procedure is called the *Restricted Open-Shell Hartree–Fock (ROHF) method*, and thus its closed-shell analogue is often abbreviated as RHF, standing for Restricted Hartree–Fock.

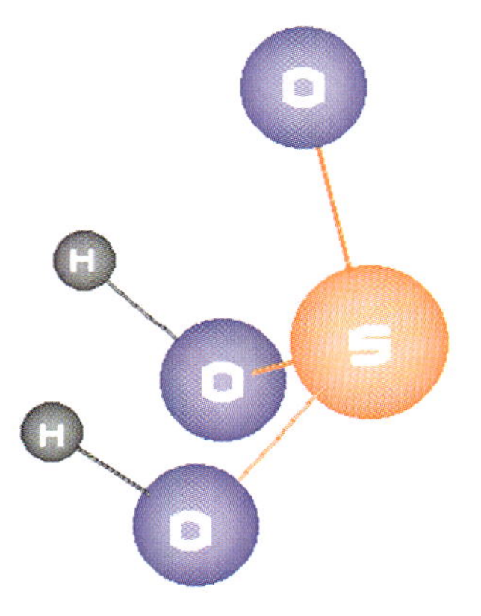

Fig. 8.1 *Ab initio* optimised geometry of the H_2SO_3 molecule.

The second method treats spin orbitals with α- and β-spins separately, thus leading to different eigenvalues and eigenvectors for α- and β-electrons. This is also illustrated in Fig. 8.2, and the calculation method is known as the *Unrestricted Open-Shell Hartree–Fock (UHF) Method*. From the different eigenvector matrices $\mathbb{C}^\alpha$ and $\mathbb{C}^\beta$ one obtains the corresponding density matrices $\mathbb{P}^\alpha$ and $\mathbb{P}^\beta$ and Fock matrices $\mathbb{F}^\alpha$ and $\mathbb{F}^\beta$. The density matrices can either be added to form a total density matrix, or subtracted, leading to the so-called spin density matrix $\mathbf{\Sigma}$:

$$\mathbb{P}^\alpha + \mathbb{P}^\beta = \mathbb{P} \quad \mathbb{P}^\alpha - \mathbb{P}^\beta = \mathbf{\Sigma}$$

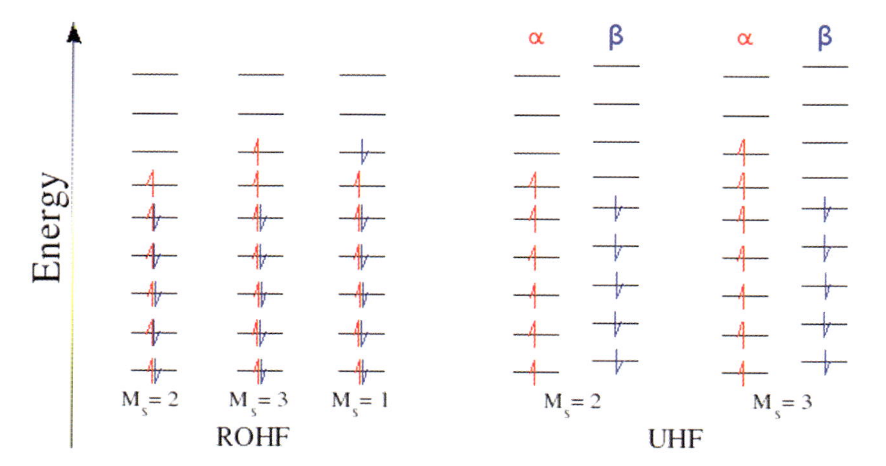

Fig. 8.2 Occupation of spin orbitals in the UHF and ROHF methods, for spin states of different multiplicity.

where Σ describes the distribution of unpaired electron density in the molecule(s) and hence is a valuable tool in the interpretation of electron spin resonance (ESR) spectra and NMR spectra of paramagnetic compounds (in the ROHF formalism, spin distribution is also evaluated, but is only based on the singly occupied spin orbitals on top of the closed-shell subsystem).

The use of two different spatial MO sets for α- and β-electrons implies that energies obtained from RHF calculations cannot be compared to UHF energies – for example, in order to estimate the relative stability of different electronic states or to calculate excitation energies. For these purposes, the ROHF formalism must be used for reasons of compatibility. The practical problem for this lies in the generally worse convergence of ROHF compared to UHF calculations.

8.1.2
Ab Initio Correlated Methods

Quantum chemical methods including electron correlation and thus – at least partially – correcting the errors of the independent-particle approximation of the Hartree–Fock formalism are called 'correlated methods'. It should be remembered that the 'correlation energy' resulting from such procedures is not a specific physical quantity, but rather just a correction of a previous approximation. It can be seen, however, as a measure indicating to what extent the correlated arrangement or re-arrangement of electron density plays a significant role for the stabilisation of a chemical system.

At this point it is helpful to recapitulate the nature of electron–electron interaction in the independent particle approach: the Coulomb integrals responsible for electron–electron repulsion are accompanied by the exchange integrals – a consequence of the antisymmetric behaviour of electrons. These exchange terms create a zone around each electron which prevents a too-close approach by another electron of the same spin. This is also called a '*Fermi hole*' surrounding the electron. For electrons with different spin, a corresponding '*Coulomb hole*' would be required, but it is not provided by the one-determinantal HF formalism. Correlated methods create this 'Coulomb hole' by providing more flexibility in the probability function, allowing the electrons to avoid unphysical vicinities in their distribution.

The common feature of all correlated *ab initio* methods is the extension from a one- to a multi-determinant probability function. These additional spin determinants are usually called 'excited states' (single/double/triple/quadruple excitations), but in fact this does not mean performing independent calculations on real excited states. The 'excited' states are constructed – as indicated before in the section on Rayleigh–Schrödinger perturbation theory – by use of the virtual (unoccupied) eigenfunctions resulting from a Hartree–Fock calculation. Therefore, use of the multi-determinant function must be seen rather as a mathematical tool to increase the flexibility of the description of the electron distribution, allowing the electrons to 'avoid' each other in a better way, rather than as a 'participation' of excited electron configurations in the ground state.

On the other hand, a simple example can demonstrate that the description of a chemical system with a single determinant is not possible in many cases. When we investigate the dissociation of a NaCl molecule in the gas phase, a single-determinantal calculation will inevitably lead to the following reaction:

$$NaCl_g \rightarrow Na_g^+ + Cl_g^-$$

This is what happens in aqueous solution under the influence of solvent and hydration, but not in the gas phase. The reason for this is that the ground state of the NaCl molecule is correctly described by a closed-shell (M = 1) determinant, and by increasing the separation of the atoms in the dissociation, the system is forced to remain in a closed-shell state, which can only be achieved by dissociation into two closed-shell ions. The correct description of the gas-phase reaction would be, however,

$$NaCl_g \rightarrow Na_g^{\cdot} + Cl_g^{\cdot}$$

that is the dissociation into atoms. Both atoms have one unpaired electron and thus represent an open-shell state, and they must be described by different determinants, namely open-shell configurations. It is obvious, therefore, that the complete reaction can only be treated by a multi-determinantal approach, allowing a continuous change in the contribution of the different determinantal functions to the total function.

There are many ways to deal with multi-determinantal approaches, the most important groups of which can be summarised as follows

- Configuration Interaction (CI)
- Multi-Configuration SCF (MCSCF)
- Coupled-Cluster (CC)
- Pair Methods
- Perturbational Methods

The common feature of all methods is the construction of other spin determinants using virtual orbitals. In the following, a very brief outline of the principal methodical differences will be given, but even this summary cannot cover the many aspects of correlated methods, which would by far exceed the character of a book introducing the basics of computational chemistry. High-level correlated calculations are still largely a subject for specialists, and for further details – in particular the mathematical formalism – readers are referred to more specialised, advanced-level books.

8.1.2.1 Configuration Interaction Methods

Configuration Interaction (CI) methods construct a set of 'excited' states, in which one, two, three, four, or more electrons are removed from occupied orbitals and are assigned to a corresponding number of virtual orbitals of the previous Hartree–Fock calculation. To specify the 'excitations', one talks of single, double, triple, quadruple ... (S,D,T,Q ...) excited spin determinants. With this set of 'excited' determinants $\{\Phi_I\}$ the CI-function $|\Psi^{CI}\rangle$ is built by linear combination according to

$$|\Psi^{CI}\rangle = |\Psi^{HF}\rangle + \sum_I C_I|\Phi_I\rangle,$$

and the linear combination coefficients are optimised with the variational ansatz to minimise the total energy. It is clear that the computational demand of such calculations depends not only on the number of basis functions used in the HF calculations (and it does not make much sense to apply correlated methods with small basis sets), but also on the number of 'excited' states considered – that is, the number of determinants $|\Phi_I\rangle$ entering the calculations. Thus, CI-SDTQ is by orders of magnitude more computer-time consuming than CI-SD. A full CI calculation including all possible 'excited' determinants is, therefore, only possible for very small molecules. For everything else one must find suitable restrictions.

The difference of energies obtainable by optimal functions $|\Psi^{CI}\rangle$ and $|\Psi^{HF}\rangle$ is called *'correlation energy'*, and it corresponds to the difference between the Born–Oppenheimer (BO) limit (for molecules) or the non-relativistic limit (for atoms) and the Hartree–Fock limit. Therefore, in order to obtain a realistic value for the correlation energy, it is necessary to perform a HF calculation that is not too far from the HF limit, and to use a sufficiently large set of determinants in the CI calculations. In all other cases the correlated methods (this is also true for the other methods discussed hereafter) will simply provide an idea about the influence of electron correlation, but not a good estimate of the correlation energy. Figure 8.3 shows the dependence of the total energy on the number of basis functions for a HF calculation of water and the corresponding curve for a calculation including electron correlation.

8.1.2.2 **Multi-Configuration Methods**

The Multi-Configuration SCF (MCSCF) method follows the same principles as the CI method, but in a more elegant fashion. Here, the linear combination coefficients c_{ij} of the spin orbitals and the coefficients C_I for the determinants are simultaneously optimised. Despite its elegance, this procedure is not widely used, mainly because of convergence problems, especially in larger systems.

8.1.2.3 **Coupled Cluster Methods**

The main idea of the Coupled Cluster (CC) method is essentially the same as for CI methods – the incorporation of electron correlation by inclusion of excited states into the probability function. The main difference lies in the construction of these excited states. The linear combination ansatz of the CI approach:

$$|\Psi^{CI}\rangle = |\Psi^{HF}\rangle + \sum_I C_I|\Phi_I\rangle$$

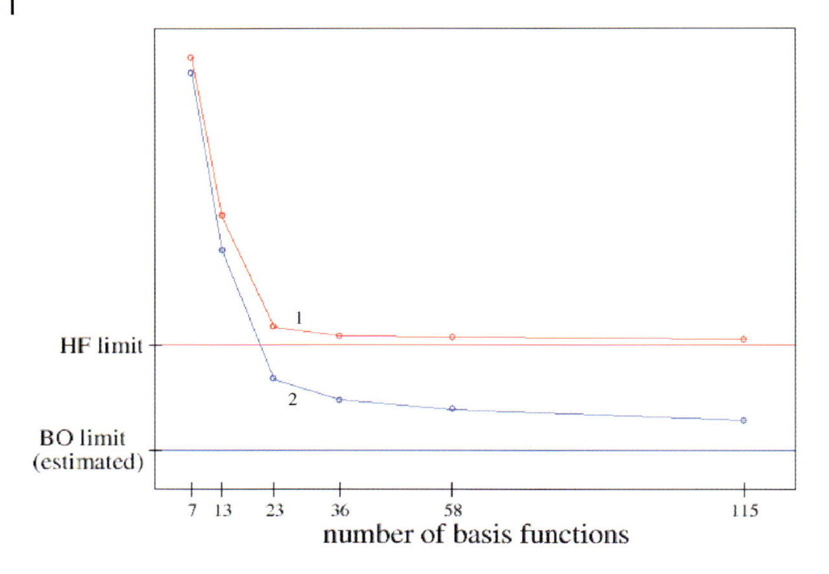

Fig. 8.3 *Ab initio* calculation of water: dependence of total energy on number of basis functions (1: HF calculation, 2: calculation including electron correlation).

can also be formulated as the action of a linear excitation operator $\hat{T}$ acting on a reference function (in this case the Hartree–Fock function)

$$|\Psi^{CI}\rangle = \hat{T}|\Psi^{HF}\rangle,$$

where $\hat{T} = \hat{T}_S + \hat{T}_D + \hat{T}_T \ldots$, referring to single (S), double (D), triple (T) ... excitations. In the CC method, an exponential excitation operator $e^{\hat{T}}$ is employed instead. In order to distinguish the method from CI, the coefficients (also called amplitudes) are symbolised by t instead of C. Thus, the total probability function results as:

$$|\Psi_{\mathrm{CC}}\rangle = e^{\hat{T}}|\Psi_{\mathrm{HF}}\rangle = \Psi_{\mathrm{HF}} + \sum_i t_{\mathrm{i}}\Psi_{\mathrm{S}} + \sum_i \sum_j t_{\mathrm{ij}}\Psi_{\mathrm{D}} + \cdots$$

$$e^{\hat{T}} = \sum_k \frac{1}{k!}\hat{T}_k = 1 + \hat{T} + \tfrac{1}{2}\hat{T}^2 + \tfrac{1}{6}\hat{T}^3 + \cdots$$

The exponential form of the excitation operators has some advantages compared to the linear one. Employing a Taylor series for the exponential representation of the excitation operator results in a sum containing – besides the excitations covered by CI – mixed terms in addition:

$$e^{\hat{T}} = 1 + \hat{T}_S + \left(\hat{T}_D + \tfrac{1}{2}\hat{T}_S^2\right) + \left(\hat{T}_T + \hat{T}_D\hat{T}_S + \tfrac{1}{6}\hat{T}_S^3\right) + \cdots$$

Therefore, the CC method seizes a larger amount of electron correlation than a CI method of the same excitation level.

Projection of the CC wavefunction onto the reference state yields the CC energy,

$$E_{\mathrm{CC}} = \langle \Psi_{\mathrm{HF}} | e^{-\hat{T}} \hat{H} e^{\hat{T}} | \Psi_{\mathrm{HF}} \rangle,$$

while projections on the excited states (S, D, T) lead to the CC equations, which allow the calculation of the amplitudes $t_i, t_{ij}, \ldots$ which are determined iteratively:

$$\langle \Psi_{\mathrm{S}} | e^{-\hat{T}} \hat{H} e^{\hat{T}} | \Psi_{\mathrm{HF}} \rangle = \emptyset$$
$$\langle \Psi_{\mathrm{D}} | e^{-\hat{T}} \hat{H} e^{\hat{T}} | \Psi_{\mathrm{HF}} \rangle = \emptyset \ldots$$

Although the CC method is very demanding in terms of computational effort due to the exponential approach, it is at present the state-of-the-art method for very accurate correlated *ab initio* calculations. Commonly, it is only employed for calculations up to a lower excitation level (S,D,T, less frequently Q). If all excitations were taken into account ('Full' approach), CC and CI would become equivalent, but as Full CC is more demanding than Full CI, the latter method becomes preferable.

8.1.2.4 Pair Methods

The pair methods have been developed not only to reduce the computational effort for the inclusion of electron correlation, but also to obtain a more illustrative description of the phenomenon. As can be recognised from the name, these methods describe correlation as occurring between pairs of electrons. This requires an assignment of the electrons to localised spin orbitals (which were discussed in Chapter 5 as a very convenient instrument to reproduce chemical bonds) and 'lone pairs' in a molecule. Working with these pairs of electrons allows a reduction of the effort to evaluate correlation contributions by topological criteria. First, one can separate these contributions into 'intra-pair' and 'inter-pair' correlations. Second, one can set a distance limit for inter-pair contributions, assuming that a mutual influence exists only in a closer neighbourhood. Starting from the *Independent Electron Pair Approximation (IEPA)*, further pair methods have been developed such as the Coupled Electron Pair Approximation (*CEPA*) and the Pair-Natural Orbital CI (*PNO-CI*). The CC methods which represent the most advanced and versatile correlated methods in practice, also have their origin in the pair method formalism.

8.1.2.5 Perturbational Methods

Last, but by no means least, the perturbational methods must be mentioned. These are the most popular methods, probably because of their simplicity, low computational cost and user-friendly implementation in many commercial programme packages. According to the perturbational method used in the standard programmes they are called *Møller-Plesset* or *MP/n* perturbational methods, where the letter *n* indicates, up to which level 'excited' determinants are considered. Thus, an MP/2-level calculation is still quite affordable, while MP/4 is already very demanding in terms of computer time and resources. The main disadvantage of the perturbational formalism, compared to the previously discussed methods, is the possibility of overestimating the correlation effect, as the practice of MP/2 calculations has repeatedly shown.

8.2 Semiempirical MO Methods

Semiempirical MO methods were introduced during the 1960s, when *ab initio* HF calculations were still prohibitively expensive for molecules with more than just a few atoms. The basic philosophy of these methods is a reduction of the computational effort by simplification, the neglect of a large part of the integrals or their replacement by experimental data, combined with a parametrisation fitting the results to *ab initio*-calculated data or experimental values.

Having witnessed such enormous progress in the speed and capacity of computers, one would assume that today these methods have become obsolete, but what has happened is rather a shift of the borderline beyond which they are employed. Large biomolecules or polymers, where *ab initio* calculations are still too demanding and where more data are needed than are available from the force field calculations (as discussed in a later section), are still a domain for semiempirical calculations. Their simplicity of use, providing fixed standard basis sets and simple outputs, may be another feature that still attracts some chemists. It should be mentioned, however, that these methods produce reasonable results only for a limited subset of compounds, mostly restricted to organic molecules which have formed the main basis for fitting the parameters and for which errors often compensate or are less critical due to the simplicity of the systems with regard to theoretical requirements.

The basic simplifications of all semiempirical MO procedures compared to *ab initio* Hartree–Fock theory are listed below and will be elaborated in the following paragraphs:

- Restriction to valence electrons
- Neglect of overlap
- Neglect of differential overlap
- Empirical approximations of core Hamiltonian elements

The first approximation – the *restriction to valence electrons* – sounds reasonable, at least from an intuitive chemical viewpoint, by assigning the 'relevant' chemistry to this region of the atoms. It can be an error source, however; for example, ions such as Na^+ are represented by a bare nuclear charge with empty atomic orbitals, and when they interact with another molecule, electron density will be transferred in too high an amount to these empty functions, thus causing artificial stabilisation effects. In all-electron calculations, polarisation of the inner electrons of the ion would have compensated for this effect. Such effects are the reason, why in *ab initio* ECP calculations the inner core region is always selected sufficiently small to retain some inner electrons within the region described by explicit functions.

The second and third approximations – *neglect of overlap* and *neglect of differential overlap* – are almost always combined, and are also known in this combined form as the *Pariser–Parr–Pople (PPP) scheme*. The neglect of overlap means that $S_{rs} = \delta_{rs}\ \forall r, s$, and thus makes the Löwdin orthonormalisation procedure preceding an *ab initio* calculation unnecessary. However, this also means that eigenvectors from *ab initio* calculations and from a semiempirical method cannot be directly compared. The full neglect of differential overlap, also called *zero differential overlap approximation (ZDO)* is characterised by

$$(rs/tu) = \gamma_{rt} \cdot \delta_{rs} \cdot \delta_{tu},$$

and actually means that all three- and four-center integrals are neglected, even a part of the two-center integrals, namely those of the type (rs/rs). γ_{rt} is a short notation for the two-center integrals (rr/tt). The neglect of a part of the two-center integrals can lead to specific problems and is corrected, therefore, in some modifications of the semiempirical methods, as we will see later.

Finally, the elements of the core Hamiltonian are approximated, differentiating between diagonal and off-diagonal elements. In the latter case, elements containing functions of the same atom are neglected in analogy to the neglect-of-overlap approximation:

$$H^c_{rs} = \emptyset \quad \forall r, s = f(A),\ r \neq s$$

If functions of different atoms are involved, the elements are first considered as consisting of two terms according to

$$H^c_{rs} = \langle r| - \tfrac{1}{2}\Delta - \hat{V}_a - \hat{V}_b|s\rangle - \sum_{C \neq A,\,B} \langle r|\hat{V}_C|s\rangle,$$

and the second term is neglected, again in analogy to the neglect-of-overlap approximation. The first term and thus the elements H^c_{rs} are empirically approximated as

$$H^c_{rs} = \beta_{rs} \cdot S_{rs},$$

where β_{rs} is a type of fitting parameter, usually composed of function-specific or atom-specific parameters according to

$$\beta_{rs} = \frac{\beta_r + \beta_s}{2} \quad \text{or}$$

$$\beta_{AB} = \frac{\beta_A + \beta_B}{2}$$

This approximation is known as the *Mulliken–Wolfsberg–Helmholtz* approach. In contemporary semiempirical methods, the parameters β are chosen to reproduce *ab initio* calculations of a larger set of molecules as well as possible, but in earlier approaches they were estimated as arithmetic or geometrical mean of the diagonal elements of the core Hamiltonian, similar to the historical *Extended Hückel calculations*:

$$H_{rs}^{c} = K \cdot S_{rs}\left(\frac{H_{rr}^{c} + H_{ss}^{c}}{2}\right) \quad \text{or:}$$

$$H_{rs}^{c} = K \cdot S_{rs}\sqrt{H_{rr}^{c} \cdot H_{ss}^{c}}$$

where K is a constant for adjusting the relations.

It might seem paradox that for the calculation of the H_{rs}^{c} terms one needs exactly those overlap integrals S_{rs}, which have been neglected in the second step of the approximations (and which must be actually calculated, therefore, in every semiempirical method!). The reason for this is the fact that overlap between functions of different atoms is an essential factor in the superposition of probabilities as we know from the discussion of the chemical bond, and has thus to be considered in some part of the framework.

The diagonal elements of the core Hamiltonian can be expressed as a sum of the kinetic energy of the electron plus the interaction with its own nucleus, and the interaction of this electron with the other nuclei of the system:

$$\begin{aligned} H_{rr}^{c} &= \langle r| - \tfrac{1}{2}\Delta - \hat{V}_{A}|r\rangle - \sum_{B \neq A} \langle r|\hat{V}_{B}|r\rangle \\ &= U_{rr} - \sum_{B \neq A} \langle r|\hat{V}_{B}|r\rangle \end{aligned}$$

The first term, U_{rr}, can be related to measured data such as ionisation energies and electron affinities.

The scheme outlined here is more or less the ‘backbone’ of all semiempirical MO methods, and we will now examine its implementation in the ‘classical’ semiempirical methods *CNDO* (complete neglect of differential overlap) and *INDO* (intermediate neglect of differential overlap).

$S_{rs} = \delta_{rs}$ and $(rs/tu) = \gamma_{rt} \cdot \delta_{rs} \cdot \delta_{tu}$ are fully implemented in the CNDO procedure, and for the ZDO approximation the following further simplifications are introduced:

$$\gamma_{rt} = \gamma_{AA} \quad \forall r, t = \varphi(A)$$

$$\gamma_{rt} = \gamma_{AB} \quad \forall r = \varphi(A),\ t = \varphi(B)$$

which means that only the *type* of nuclei, and not the *specific basis functions* is considered in the expressions for the two-electron integrals. This reduces the very large list of these integrals to a simple symmetric matrix of the dimension N, where N is the number of nuclei in the system, and represents (besides the restriction to valence electrons) the main saving of computer time in semiempirical calculations.

The diagonal elements H_{rr}^c are approximated by introducing

$$-\tfrac{1}{2}(I_r + A_r) = U_{rr} + \left(Z_A - \tfrac{1}{2}\right)\gamma_{AA},$$

where I_r is the ionisation energy and A_r the electron affinity assigned to an electron function φ_r. The second term is an average which is supposed to take care of the influence on the electron–electron repulsion when removing or adding an electron from/to the atom. For the interaction of the electron in φ_r with the other nuclei the expression $\langle r|\hat{V}_B|r\rangle$, abbreviated as V_{AB} is replaced by

$$V_{AB} = Z_B \cdot \gamma_{AB},$$

as this compensates to some extent the neglect of so many electron–electron repulsion terms in the ZDO formalism.

The off-diagonal elements H_{rs}^c are neglected for functions of the same atom, otherwise treated as the following:

$$H_{rs}^c = \beta_{AB}^\circ \cdot S_{rs} \quad \forall r = \varphi(A),\ s = \varphi(B)$$

$$\beta_{AB}^\circ = k \cdot \frac{\beta_A^\circ + \beta_B^\circ}{2}$$

where k is a constant and β_A° and β_B° are atom-specific parameters. We can now compare the *ab initio* Fock matrix elements

$$F_{rs} = H_{rs}^c + \sum_{t,u} P_{tu}\left[(rs/tu) - \tfrac{1}{2}(rt/su)\right]$$

with their CNDO analogues. For the off-diagonal elements we obtain

$$F_{rs} = H_{rs}^c - \tfrac{1}{2} P_{rs}(rr/ss)$$

and further

$$F_{rs} = \beta_{AB}^0 S_{rs} - \tfrac{1}{2} P_{rs}\gamma_{AB}$$

The diagonal elements can be successively developed

$$F_{rr} = H_{rr}^c + \sum_t P_{tt}\left[(rr/tt) - \tfrac{1}{2}(rr/rr)\right]$$

$$= H_{rr}^c + \sum_B P_{BB}\gamma_{AB} - \tfrac{1}{2}P_{rr}\gamma_{AA}$$

$$F_{rr} = H_{rr}^c + \left(P_{AA} - \tfrac{1}{2}P_{rr}\right)\gamma_{AA} + \sum_{B \neq A} P_{BB}\gamma_{AB}$$

$$H_{rr}^c = U_{rr} - \sum_{B \neq A} V_{AB}$$

$$F_{rr} = U_{rr} + \left(P_{AA} - \tfrac{1}{2}p_{rr}\right)\gamma_{AA} + \sum_{B \neq A}(P_{BB}\gamma_{AB} - V_{AB})$$

$$\text{with } V_{AB} = Z_B \cdot \gamma_{AB} \quad \text{and} \quad -\tfrac{1}{2}(I_r + A_r) = U_{rr} + \left(Z_A - \tfrac{1}{2}\right)\gamma_{AA}$$

finally resulting as

$$F_{rr} = -\tfrac{1}{2}(I_r + A_r) + \left[(P_{AA} - Z_A) - \tfrac{1}{2}(P_{rr} - 1)\right]\gamma_{AA} + \sum_{B \neq A}(P_{BB} - Z_B)\gamma_{AB}$$

We see that these elements are basically determined by the empirical (experimental) values for ionisation energy and electron affinity, and that electron–electron interactions play an increasing role, when redistribution of electron density becomes larger – that is, when $P_{rr} \neq 1$ and $P_{AA} \neq Z_A$ and $P_{BB} \neq Z_B$. Due to the neglect of overlap the sum of all diagonal elements P_{rr} delivers the Mulliken electron population of an atom

$$\sum_{r \in A} P_{rr} = P_{AA}$$

and $Z_A - P_{AA}$ the partial charge on this atom according to Mulliken.

For the off-diagonal elements of the CNDO Fock matrix the overlap between functions of different atoms determines their size and thus S_{rs} remains a key player in all 'neglect-of-overlap' methods.

One important feature of semiempirical procedures is that standardised basis sets are assigned to each atom within the program. Usually, these are single-valence STO basis sets, which makes them inflexible on the one hand, but easy to use by the inexperienced on the other hand.

The SCF procedure in CNDO and all other semiempirical MO methods remains the same as in *ab initio* Hartree–Fock calculations, the starting solutions for $\mathbb{C}^\circ$ being obtained by diagonalisation of $\mathbb{H}^c$. Due to the restrictions to valence electrons and the limited number of basis functions, the semiempirical Fock matrices have a much smaller dimension than their *ab initio* analogues, and the addition of two-electron electron integrals to the F_{rs} elements takes almost no time at all, thus making the SCF procedure considerably less time-consuming.

The *INDO* formalism, which forms the basis of many further semiempirical procedures, differs from the CNDO scheme mostly by modifications in the neglect of differential overlap. In particular, the neglect of (rt/rt) integrals, while (rr/tt) integrals are retained, can be problematic, notably when comparing different electronic states of excited molecules in open-shell configurations. INDO corrects this problem by considering all types of two-center integrals in the form of parameters included in the program code, the *Slater–Condon parameters*. Further, the terms U_{rr} are evaluated taking into account the type of the function φ_r.

Starting from INDO, numerous other semiempirical methods have been developed, mostly with different and/or additional parametrisations. The first of these was MINDO/3, followed by others such as SINDO and SPINDO. At present, the most common methods are the 'Austin Model' AM/1 and the PM/3 method. For details of these methods, the reader is referred to specialised books and the original literature cited therein.

As already mentioned, semiempirical MO methods are no longer widely used, and for good reasons their use is restricted to large organic molecules, although parametrisations have been made for all elements up to krypton. Organic molecules have always been seen as a type of 'primitive subset' among chemical compounds for theoretical treatment because of the limited types of atoms involved and the low atomic number of these elements. Whenever a more 'complicated' atom (e.g., a central metal atom in an enzyme) is involved, this simplicity and the related fortunate error compensations disappear, and one must revert to more sophisticated calculation methods.

Multi-determinantal approaches have also been developed for semiempirical MO procedures, such as CNDO-CI or PCILO. These can be seen as mere attempts to improve the description of the systems, but they cannot give real indications towards correlation energy contributions, as the reference energy is not related to a Hartree–Fock energy or limit. On the other hand, the use of experimental parameters in the formalism implies that 'correlated' data are a part of the method, thus adding some (uncontrollable) correlation contribution to the results. This also explains, why semiempirical results sometimes gave 'better' results for structure optimisations than one-determinantal *ab initio* methods, (e.g., for the molecular ion CH_5^+).

8.3 Density Functional Methods

The basic philosophy of density function theory (DFT) – namely to use the electron density instead of a 'wave function' to obtain information about chemical systems – originated almost simultaneously with Schrödinger's equation (1926) through the studies of *Thomas* and *Fermi* in 1927. This approach was taken up again during the early 1950s by *Slater*, who sought a simpler form for the complicated HF exchange term by defining the exchange energy as

$$E_X = \tfrac{1}{2}\iint \frac{\rho(r_1)\hat{h}_X(r_1;r_2)}{r_{12}}\,dr_1\,dr_2$$

as a function of the electron density and a local exchange operator. As it had been shown that the 'Fermi hole' created by the antisymmetry corresponds to a sphere containing one elementary charge, with the so-called *Wigner–Seitz* radius

$$r_S = \left(\frac{3}{4\pi}\right)^{\frac{1}{3}} \rho(r_1)^{-\frac{1}{3}},$$

it was possible to simplify the expression for the exchange energy to

$$E_X[\rho] \cong C_X \int \rho(r_1)^{\frac{4}{3}}\,dr_1,$$

where C_X is a numerical constant. In order to improve this approximation, an adjustable parameter α was introduced, leading to the *Hartree–Fock–Slater (HFS) X α* method:

$$E_{X\alpha}[\rho] = -\tfrac{9}{8}\left(\frac{3}{\pi}\right)^{\frac{1}{3}} \alpha \int \rho(r_1)^{\frac{4}{3}}\,dr_1$$

The application of this method to solid-state model compounds seemed promising in some aspects, but its use for 'ordinary' molecules was very disappointing, and consequently the once quite popular method is resting now in the graveyard of outdated semiempirical procedures.

Modern DFT began during the 1960s with the formulations of *Hohenberg* and *Kohn*. Their first theorem states that the many-particle ground state is a unique functional of $\rho(r)$, as the external potential – determining the Hamiltonian – is a unique functional of $\rho(r)$. Therefore, all components of the ground state energy – that is, kinetic energy, electron–electron interactions and electron–nuclei interactions – can also be expressed as functional of the electron density:

$$E_\circ[\rho_\circ] = T[\rho_\circ] + E_{ee}[\rho_\circ] + E_{Ne}[\rho_\circ]$$

The second Hohenberg–Kohn theorem introduces the variational principle into this approach, stating that the functional F_{HK} which delivers the true ground state energy must correspond to the 'true' ground state density $\rho_\circ$. Any other trial density $\tilde{\rho}$ leads to higher energy values:

$$E_\circ \leq E[\tilde{\rho}] = T[\tilde{\rho}] + E_{Ne}[\tilde{\rho}] + E_{ee}[\tilde{\rho}]$$

This provides the basis for a variational optimisation of the density functional. It should be noted that this is a formalism for the ground state that cannot be extended in straightforward manner to excited states!

By these theorems, we know that DFT should work in principle, but we do not know yet how to construct a functional that correctly reproduces the ground state energy. A practical solution to this problem was introduced by *Kohn* and *Sham*, by an approach incorporating the works of *Thomas*, *Fermi* and *Slater*, as well as the orbital concept well-known from *Hartree–Fock* theory. With the use of orbitals and a DFT-adapted form of a Slater determinant, one can define the Kohn–Sham equations with a one-electron operator, similar to the HF equations as:

$$\hat{f}_{KS}|\psi_i\rangle = \varepsilon_i|\psi_i\rangle$$
$$\hat{f}_{KS} = -\tfrac{1}{2}\Delta + \hat{V}_S(r)$$

and the electron density can be obtained from the $\psi_i^* \psi_i$ values.

In contrast to HF theory, the exact kinetic energy cannot be computed from a functional, and the approach to evaluate this energy in analogy to the HF formula

$$T_S = -\tfrac{1}{2}\sum_i^N \langle\psi_i|\Delta|\psi_i\rangle$$

is not equivalent to the latter, as in DFT we are dealing with 'non-interacting' electrons. Therefore, the functional $F[\rho]$ was separated into

$$F[\rho(r)] = T_S[\rho(r)] + J[\rho(r)] + E_{XC}[\rho(r)]$$

This equation contains, besides the kinetic energy term and Coulomb term, the so-called exchange-correlation energy term E_{XC}, which is defined by

$$E_{XC}[\rho] \equiv (T[\rho] - T_S[\rho]) + (E_{ee}[\rho] - J[\rho]) = T_C[\rho] + E_{ncl}[\rho],$$

and which contains not only (as its name suggests) exchange and correlation effects, but also a correction term for the kinetic energy which is not covered by T_S (T_C stands for the 'true' kinetic energy, E_{ncl} summarises all non-classical effects, i.e. self-interaction correction, exchange and correlation).

An important task to be solved is the formulation of the potential operator $\hat{V}_S$ in the Kohn–Sham operator $\hat{f}_{KS}$. This is done by approximating it as an effective operator $\hat{V}_{eff}$, defined by

$$\hat{V}_{eff} = \int \frac{\rho_{r_2}}{r_{12}} dr_2 - \sum_A \frac{Z_A}{r_{iA}} + V_{XC}$$

in which the potential V_{XC} describing the exchange-correlation energy effect is unknown and can only be defined as a functional derivative of the exchange-correlation energy:

$$V_{XC} \equiv \frac{\delta E_{XC}}{\delta \rho}$$

If E_{XC} and V_{XC} were exactly known, one could indeed obtain the exact ground state energy. However, as this is not the case, one must choose some way of approximations, and this is where the problems of density functional methods begin.

Although many parts of the Kohn–Sham formalism appear very similar to HF theory, there are some substantial differences, which can be understood by a brief summary of the Kohn–Sham procedure. The method begins from a reference system of N *non-interacting* particles, defined by a Slater determinant which should represent the exact ground state by producing – via the KS orbitals contained in the determinant – a density $\rho_S = \rho_\circ$. The actual density ρ_S is included in the expression for the energy. The KS orbitals are the solutions of N eigenvalue equations of the 1-particle operator $\hat{f}_{KS}$, which delivers the non-interacting kinetic energy, and whose potential energy operator must be chosen to fulfil the condition $\rho_S = \rho_\circ$. V_S contains, besides the interaction of the electron with the nuclei and the classical Coulomb interaction, the (unknown) term V_{XC}, which can be (iteratively via the functions $|\psi_i\rangle$) obtained according to $\frac{\delta E_{XC}}{\delta \rho}$. V_{XC} has to take care of the remaining kinetic energy contribution and the non-classical exchange and correlation effects. The KS orbitals have – as the HF spin orbitals – no real physical meaning, but are considered by some authors also as a practical tool for qualitative MO models. However, Koopmans' theorem is not applicable to the eigenvalues ε_i of the KS orbitals.

In a somewhat sloppy way one could also define the main difference between HF and DF approach by stating that in HF we work with exactly defined operators and optimise our results by improving the probability function, whereas in DFT methods we manipulate the operator (with the help of some functions to calculate ρ, albeit in the KS formalism) to obtain – via the electron density – the (ground state) energy, without the need to obtain a physically significant probability function.

As the overall density employed in this formalism does not contain any information about the spin state of the system, it cannot provide the flexibility of the HF formalism to deal with open-shell systems. If one needs to account for these systems, then functionals must be introduced explicitly, depending on α and β spin densities, and this leads to the Unrestricted Kohn–Sham (UKS) formalism.

Another difficulty of the KS method arises when dealing with excited states. A number of promising proposals to master the problem that the density functional refers only to the ground state have been made, for example by multi-determinant approaches or via *time-dependent density functional theory (TDDFT)*, in which a time-dependent perturbation produces the correlated excitation energies of the unperturbed system. The latter method has already been implemented in commercial quantum chemical program packages, and can be considered, therefore, as the most successful approach in this context.

The most crucial problem of DFT remains, however, the search for suitable approximations for V_{XC} and E_{XC}, respectively. The main types of approximation can be classified as local density, generalised gradient, and hybrid functional approximations (increasing in accuracy and suitability for molecular systems in this order), and these are briefly introduced in the following paragraphs.

8.3.1
Local Density Approximation (LDA)

This approximation has its roots in the concept of a homogeneous electron gas – that is, the distribution of all electrons above a positively charged background (such a model was briefly mentioned while discussing the classical ideas of chemical bonding in the case of the metallic bond). It is quite clear that LDA might be an acceptable approximation for simple metals, but would never provide a suitable model for molecules with their very specific distribution of electron density. The fact that the homogeneous electron gas is the only case for which one can define the exchange and correlation terms with high accuracy, is the only reason why LDA still plays such an important role in DFT. The basic formulation for E_{XC} in the LDA is

$$E_{XC}^{LDA}[\rho] = \int \rho(r)\varepsilon_{XC}(\rho(r))\,dr$$

where $\varepsilon_{XC}(\rho(r))$ is the exchange-correlation energy per particle of the uniform electron gas, which can be decomposed as

$$\varepsilon_{XC}(\rho(r)) = \varepsilon_X(\rho(r)) + \varepsilon_C(\rho(r))$$

to exchange and correlation part. According to an approximation by Slater, ε_X can be written as

$$\varepsilon_X = -\frac{3}{4}\sqrt[3]{\frac{3\rho(r)}{\pi}}.$$

For ε_C, numerous expressions have been proposed, and they are usually characterised by the initials of their authors, plus a number for the respective variant or publication year of this functional and its specific parametrisation (e.g. VWN5 or PW92).

By extending the LDA to the unrestricted case, one arrives at the *Local Spin-Density Approximation* (*LSD*, not to be mistaken for an hallucinogenic drug!):

$$E_{XC}^{LSD}[\rho_\alpha, \rho_\beta] = \int \rho(r)\varepsilon_{XC}(\rho_\alpha(r), \rho_\beta(r))\,dr$$

from which we can derive relations for E_{XC} for either spin-compensated systems or systems with uneven spin distribution, also called 'spin-polarised'.

As mentioned above, LDA and LSD seemed suitable for some solid-state compounds, but are a very inaccurate description for molecules, and further improvements appeared mandatory, before using DFT in chemistry. These will be introduced subsequently.

8.3.2 Generalised Gradient Approximation (GGA)

By supplementing the density $\rho(r)$ by its gradient $\nabla\rho(r)$, one can introduce the non-homogeneity of the electron density distribution. The LDA is seen as the first term of a Taylor expansion, which is extended by the next lowest term, leading to the *Gradient Expansion Approximation (GEA)*:

$$E_{XC}^{GEA}[\rho_\alpha, \rho_\beta) = \int \rho\varepsilon_{XC}(\rho_\alpha, \rho_\beta)\, dr + \sum_{\sigma,\sigma'} \int C_{XC}^{\sigma,\sigma'}(\rho_\alpha, \rho_\beta) \frac{\nabla\rho_\sigma}{\rho_\sigma^{\frac{2}{3}}} \frac{\nabla\rho_{\sigma'}}{\rho_{\sigma'}^{\frac{2}{3}}}\, dr + \cdots$$

This functional form has not yet led to any real improvement, mostly because it does not properly describe the exchange term. By forcing restrictions valid for the 'real' exchange behaviour by truncating the summation and setting parts of the expression to zero, it is possible to obtain the *Generalised Gradient Approximation (GGA)*:

$$E_{XC}^{GGA}[\rho_\alpha, \rho_\beta] = \int F(\rho_\alpha(r), \rho_\beta(r), \nabla\rho_\alpha(r), \nabla\rho_\beta(r))\, dr$$

E_{XC}^{GGA} can be split again to E_X^{GGA} and E_C^{GGA}, to find separate solutions for these two terms.

The functional F is of central importance in this approximation, and again a large number of proposals exist, mostly obtained by trial-and-error or fitting procedures, with the corresponding functionals again being named after their constructors (e.g., FT97, PW91, PBE, BLYP, etc.).

8.3.3 Hybrid Functionals

As the solutions for the exchange part of E_{XC} have always been very approximate and error-prone (including the problem of unphysical self-interaction), it appeared most appropriate to obtain this part (i.e., the 'Fermi hole') from the exact (i.e., the Hartree–Fock) formalism, and to formulate a functional only for the correlation part (i.e., a 'Coulomb hole') based on the Kohn–Sham approach – which is missing in one-determinantal HF:

$$E_{XC} = E_X^{exact} + E_C^{KS}$$

As any method implementing this separation requires, besides the density functional approach, also a HF step in the calculations, the terms *'hybrid functional methods'* or *'HDFT'* have been adopted. Further, the usual constructions of the adapted correlation functionals also involve the use of previously mentioned LSD and GGA functionals, thus underlining the hybrid character. The literature contains a large number of such functionals, but here we will mention only the most popular, *B3LYP* (Becke-3, Lee, Yang, Parr):

$$E_{XC}^{B3LYP} = (1-a)E_X^{LSD} + aE_{XC}^{\lambda=\emptyset} + bE_X + cE_C^{LYP} + (1-c)E_C^{LSD}$$

where $\lambda = \emptyset$ indicates the value obtained for non-interacting particles, and a,b,c are parameters determined by fitting the functional to ionisation energies, total energies and proton affinities of a set of molecules.

So far, hybrid functionals supply the best values for molecular calculations among all DFT methods, and this has made them the most popular DFT applications in practical computational chemistry.

This short description of the common DFT calculations should have clarified the fact that these methods (which also are implemented in commercial packages such as GAUSSIAN and TURBOMOLE) must be classified as semiempirical rather than as *ab initio* methods, even if a subset of computational chemists prefer their consideration as 'non-empirical' or '*ab initio*' procedures. Adjustable parameters, manifold approximations and manipulations of operator expressions, fitting procedures and the mushrooming of functionals with acronyms are typical of semiempirical quantum chemistry.

There are promising new approaches available, however, which seem to pave the way to true *ab initio DFT* by using (instead of guessing functionals) orbital-dependent expressions which correctly describe the physical behaviour of the electrons and then convert the non-local potential into a local form by using the *optimised effective potential (OEP)* strategy. This treatment also cancels the incorrect self-interaction of standard DFT methods. The introduction of an appropriate functional for the correlation term in this approach could indeed lead to a one-electron formalism which contains – in contrast to the HF analogue – also electron correlation. Due to the use of HF-like expressions and orbitals, this methodology can be considered as a non-empirical hybrid method, situated between DFT and 'wave function' theory. Encouraging results have been obtained using this approach, for example in the treatment of weak van der Waals interactions, which are not accessible to conventional DFT methods, although the method is still far from being ready as a routine procedure accessible for a larger numbers of chemists.

The established approximate DFT methods have certainly become an important new instrument for quantum chemical calculations (although not for all chemical systems!), and will thus remain an important facet of computational chemistry.

However, to consider them as an alternative superior to *ab initio* HF methods seems inadequate, at least in their present form.

Finally, at this point the often-heard argument that (approximate) DFT calculations have been performed to '... investigate the influence of correlation energy in a system' must be critically assessed. Although it is true that DFT energies contain some correlation energy, this contribution is an uncontrollable quantity (as in semiempirical MO calculations, where it enters through the experimental data), and it can be both under- and overestimated. There are numerous practical examples available which prove both cases, and ultimately one must resort to correlated *ab initio* methods if a reliable estimation of correlation energy is needed.

Test Questions Related to this Chapter

1. What is the reason for the occurrence of a basis set superposition error?
2. Which types of basis sets are recommended for heavy atoms?
3. What is the difference between ROHF and UHF calculations?
4. What is the reason for the appearance of correlation energy, and how can it be considered in quantum chemical calculations?
5. How are the 'excited states' constructed in correlated methods?
6. What are the main simplifications of semiempirical MO-SCF calculations, compared to *ab initio* calculations?
7. Why do we have to calculate the overlap integral matrix in semiempirical MO methods, despite one of the main approximations being the neglect of overlap?
8. What is the underlying principle of density functional theory, and what is its main problem in practice?
9. What are the reasons for the successive development of LDA, GGA, and hybrid functionals?
10. Why should standard DFT methods be characterised as semiempirical and not as *ab initio* calculation procedures?

9
Force Field Methods and Molecular Modelling

Not everything that is countable counts, and not everything that counts is countable.

Albert Einstein, 20th century

The methods presented in this chapter are of almost purely empirical nature, without any quantum chemical background and formalism. Nonetheless, they have gained enormous importance in theoretical and computational chemistry, as they provide an efficient and useful basis not only for theoretical investigations of large biomolecules such as RNA, DNA and proteins, but also of synthetic organic polymers and resins. Indeed, the increasing use of these methods in industrial chemistry has led to them becoming so important that today they should be considered as much a part of a chemistry curriculum as basic quantum theory and quantum chemical calculations.

It has been mentioned above that the main domain of these methods are the biopolymers, which are not yet accessible for a computational treatment by any other approach. There are however serious restrictions in the applicability of force field methods, primarily their restriction to organic molecules. Even there, however, the empirical nature of the method requires continuous control of the quality of the results, and hence one can see new, improved parametrisations to develop within relatively short time intervals. The variety of force fields available also documents that different chemical systems may require different parametrisations, and that no universal solution exists for the description of molecules by force fields. This is also the reason why the active centres of macromolecules must often be modelled by more accurate methods based on quantum chemistry, thus leading to hybrid procedures.

9.1
Empirical Force Fields

Empirical force fields consist of a large number of parameters and potential functions, a part of them following the scheme of typical molecular vibrations such as stretching and bending modes, torsional, and out-of-plane motions. Other parameters and functions refer to electrostatic interactions and to van der Waals forces,

The Basics of Theoretical and Computational Chemistry. Edited by B. M. Rode, T. S. Hofer and M. D. Kugler

ISBN: 978-3-527-31773-8

also termed 'non-bonding' interactions. The large number of different force fields developed does not allow a detailed discussion of the differences and specific strengths and weaknesses of them within the framework of this book. Therefore, only the basic principles will be considered, by referring to one of the most common force fields, AMBER – which is an acronym for 'Advanced Model Building with Energy Refinement'.

The basic terms implemented in AMBER (and in a similar form in most other force fields) refer to the bonds (stretching forces), angles (bending forces), dihedral angles (torsional forces) and to the van der Waals (R^{-12}, R^{-6}) and Coulombic $\left(\frac{1}{R}\right)$ terms:

$$E_{\text{total}} = \sum_{\text{bonds}} K_r(r - r_{\text{eq}})^2 + \sum_{\text{angles}} K_\theta(\theta - \theta_{\text{eq}})^2 + \sum_{\text{dihedrals}} \frac{V_n}{2}[1 + \cos(n\varphi - \gamma)] + \sum_i \sum_{<j} \left[\frac{A_{ij}}{R_{ij}^{12}} - \frac{B_{ij}}{R_{ij}^6} + \frac{q_i q_j}{\varepsilon R_{ij}}\right]$$

This approach calculates the energy contributions of the first three terms on the basis of the harmonic approximation, while more sophisticated force fields contain higher terms in order to include the anharmonicity of molecular vibrations. Similarly, additional terms can be found for the 'non-bonding' interactions, namely the electrostatic and van der Waals contributions.

Modern force fields contain parameters for almost the whole periodic system (which does not mean that one should attempt to model all kinds of compounds!). For each of the elements – and in particular for the elements of which organic molecules consist – a variety of different parameters are needed, depending on the chemical surrounding and binding state of the particular atom. Thus, a carbon atom has different parameters if it is situated in a saturated or unsaturated hydrocarbon, bound to oxygen in an aldehyde, ketone, ether, alcohol or carbonic acids, or to nitrogen in an amine, amide or nitrile, to name just a few examples. From this one can easily deduce that even the most common elements of biopolymers require a very large set of parameters, and that the fitting of a complete force field is a very tedious and delicate procedure. On the other hand, the large number of parameters does not allow the easy prediction of possible error sources or a specific reason for some failures – only experience collected through practical use will enable a fairly good judgement of the reliability and limitations of a given force field calculation.

9.2 Molecular Modelling Programs

The term 'Molecular Modelling' is not always used in a unique way – sometimes, it refers to highly accurate quantum chemical calculations, and sometimes just to very simple force-field methods. The most general definition would be to call any computational treatment of a chemical system 'Molecular Modelling', but within the context of this book it will be mainly used for the computer-assisted procedures to build, optimise, and interact molecules on the basis of force fields.

A number of modelling programs are available commercially from simple desktop versions such as HYPERCHEM to large and sophisticated program systems such as SYBYL, implementing various force fields as choice.

The main functions offered by these program packages are:

- *Building of molecules* in graphical mode from an atom database.
- *Optimisation of the structure* of molecules based on force field(s) and energy optimisation algorithms such as Newton–Raphson or Steepest Descent.
- *'Fitting' of molecules*, which means a quantitative comparison of the similarity of molecular (sub)structures.
- *'Docking' of molecules*, which means a quantitative investigation of the interaction of two molecules on the basis of electrostatic and van der Waals forces, including continuous adaptation of the geometry of both reaction partners (the changes of geometry also have an influence on the total energy, and thus all intra-molecular contributions must be continuously re-calculated).
- *Saving of geometrical data* after optimisation and/or docking procedures. This will not only deliver internal coordinates (bond lengths, bond angles, dihedral angles), but also produce data files suitable for subsequent quantum chemical calculations. Some modelling programs are capable of performing such calculations within an internal limited and standardised framework, or they contain interfaces to standard quantum chemical program packages.
- Performance of simple statistical simulations of Monte Carlo and/or molecular dynamics type. These simulations are discussed in more depth and detail in Chapter 10.

While the first three items in the list are usually quite fast and straightforward processes, docking and the performance of statistical simulations are much more computer-intensive and require considerable experience and skill, if they are to lead to meaningful results. We will deal with these tasks in more detail, therefore, in the subsequent section and in Chapter 10, respectively.

9.3 Docking

Let us consider the docking of an active substance with a large protein substrate, as shown in Fig. 9.1. The complicated structure of the protein is clearly the first

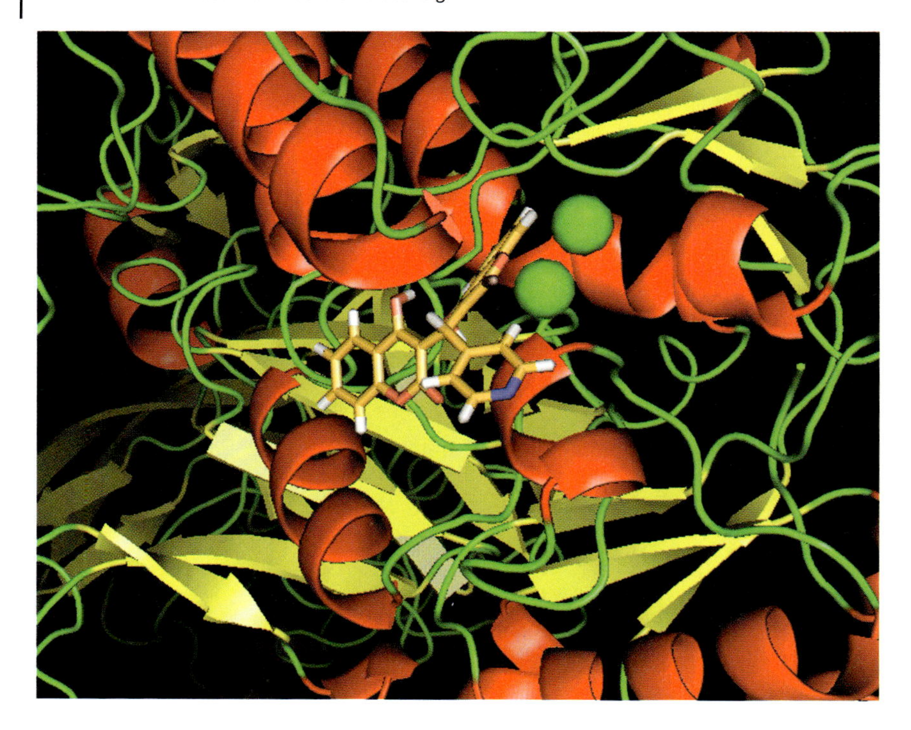

Fig. 9.1 Docking of a bis-coumarin derivative to *Bacillus pasteurii* urease. Bis-coumarins are potent urease inhibitors that bind to the active site of the enzyme. The colours of the receptor are as follows: α-helices, red; β-sheets, yellow; loops, green; the ligand is coloured according to atom types (C, orange; O, red; N, blue).

problem for this task. To date, modelling programs are not yet able to predict, from its amino acid sequence alone, the complete structure including the folding of a protein, as visible with the alpha-helices (red) and beta-sheets (yellow) in Fig. 9.1. One must start from a known structure of the same or a similar protein, and continue modelling from this starting solution: this is called *homology modelling*. The problem of this approach is the way that experimental protein structures have been determined. If they are X-ray structures, they have been obtained for a crystallised protein, which might not have the same structure as is prevalent in a biological environment (thus, such data are sometimes called necrochemical instead of biochemical data). If they have been determined in solution by two-dimensional NMR, usually very simple models are used to interpret the NMR data by a fitting procedure. In both cases, therefore, error sources are present, which might lead to a not-so-suitable starting solution for the molecular modelling and docking processes. On the other hand, the modelling process enables an optimisation of the starting geometry of the protein. In order to obtain realistic structures for the biological environment in this optimisation, the inclusion of solvent is mandatory, however, and the embedding of the protein in a sufficient amount of water molecules (tens of thousands!) considerably increases the computational effort of the modelling.

The next main task is to find possible docking places for the reaction partner (e.g., a pharmacophore) on the surface or even inside the biopolymer, and in particular the global minimum in terms of binding energy. Due to the large number of possible orientations and binding sites, an economic search process must be implemented. This procedure starts by scanning positions on a relatively wide grid at elevated temperature, moving the smaller molecule in the surrounding of the macromolecule, and accepting or rejecting the moves by a Monte Carlo-type algorithm (details of the Monte Carlo method are presented in Chapter 10). Temperature and step-width are then gradually decreased until the desired value (e.g., room or physiological temperature) and a fine grid allowing a precise localisation of the docking molecule are reached. This procedure is called '*simulated annealing*'.

One of the main problems in docking is the possibility of becoming stuck in local minima instead of proceeding to the global minimum. Repetition of the search process from different starting geometries and re-elevation of temperature are testing possibilities to determine the validity of the assumed optimal structure. Finally, the importance of solvent and ions in the system must be stressed, as they can strongly influence structure and interaction of the docking partners.

Another example of docking is illustrated in Fig. 9.2, which shows two snapshots from a dynamic docking simulation of the antimalarial drug artemisine and the haem centre of erythrocytes. In this case, docking was investigated using a molecular dynamics procedure (see Chapter 10) and the reaction partners were embedded in water. Due to the iron atom playing a central role in this process, a force field with specific parameters had to be employed in order to describe properly the intermolecular interactions.

9.4 Quantitative Structure–Activity Relationships (QSARs)

QSARs date back to a much earlier time than modelling, when chemists tried quantitatively to relate the physical properties of molecules (including structural elements) to their observed chemical, pharmaceutical or biological properties and activities. The predictor variables were distribution coefficients between polar and apolar solvents, dipole moments, and steric factors assigned to substituents, and the underlying hypothesis for the prediction of certain activities was that these properties and the structures of the active molecule and its receptor would determine the actual degree of activity.

When molecular modelling methods began to enable a much more exact determination of the topology of the active compounds, QSAR was placed on a much more accurate basis. Moreover, if the receptor's structure was also known, docking methods opened a wide field of applications to investigate potential reaction mechanisms. This also allowed one systematically to describe and compute molecular data that helped to construct QSARs and thus to predict the activity of

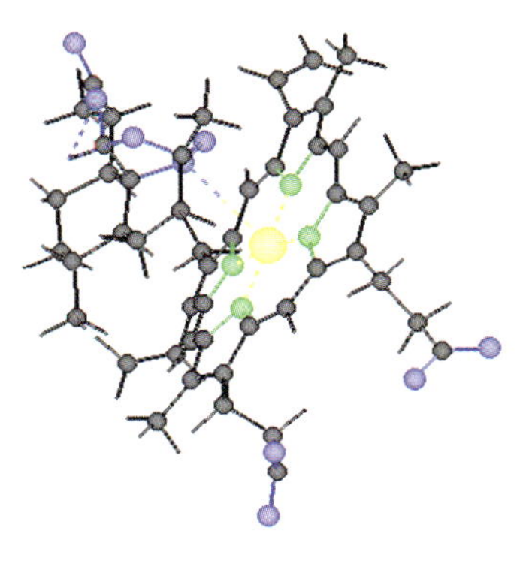

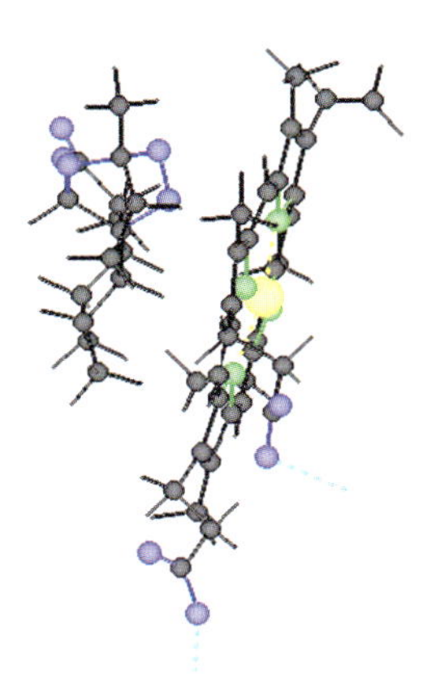

Fig. 9.2 Model docking of the antimalarial compound artemisinine to the haem receptor in erythrocytes.

new, hitherto experimentally uninvestigated compounds of the same class of substances. This new field of computational chemistry – which in the case of pharmaceutical applications is called *computer-assisted drug design* – soon gained much interest, not only for basic research but also for commercial applications in the pharmaceutical industry. Subsequently, this field has expanded to other than pharmaceutically relevant compounds, and today one speaks of *computer-assisted material design*. For organic compounds, force field modelling is a very suitable approach for such studies, but as we can see from the example given below, even in these cases the employment of quantum chemical methods can be a useful and sometimes recommended procedure.

A typical QSAR study will model a series of compounds of the same class (or with great similarity) with different substituents and determine the best docking place for the receptor, evaluating then the 'non-bonding' forces acting between the reaction partners. The term 'non-bonding' forces is somehow misleading, as one can immediately see that electrostatic and van der Waals forces lead to concrete binding energies, which can be taken as a predictor variable for the investigated activity. For a given series of compounds with known activities one can then try to establish, by statistical methods, a quantitative relationship between the calculated predictor variables and the measured activity. The relationship is usually tested not only for its quality of fitting through *regression* coefficients, but also for its '*predictability*'. For this purpose one can make use of *cross-validation* procedures; an example here is the 'leave-one-out' method, where for a series of n compounds a relationship is fitted to an equation using $(n-1)$ of these compounds, which is then used to predict the activity of the nth compound. Repeating this procedure with all n compounds leads to n deviations ΔA between predicted and actual activities, and $\sum_n (\Delta A)^2$ is a measure for the ability of the relationship to predict an unknown compound. The predictability if often normalised to 1 – that is, a fully correct prediction corresponds to a q^2 value of 1, and the lower the value is, the more fallible become the predictions.

In many cases, where QSAR is desired, the receptor is not known. In this case – and in all cases of quantitative structure–property relationships, where no receptor has to be considered – it is possible to establish similar equations for the prediction of activities/properties, based on physical and chemical data obtained for a series of compounds from either experimental or theoretical methods. In this context, quantum chemical calculations of molecular properties have acquired a specific importance, as they enable a unified and rapid access to the basis for chemical reactivity, the electron distribution within a molecule. In practice, it is not the specific values of $\Psi^*\Psi$ that are utilised to describe electron distribution, but rather the practical (albeit strictly non-observable) quantities such as the Mulliken atomic populations. The establishment of quantitative relationships between atom-focused data and activities or properties also facilitates identification of the active centre of the compounds studied. To illustrate the general set-up of QSAR investigations and the use of computational chemistry results in such studies, an example is given below in which a relationship between electron density distribution and pharmaceutical activity is sought, by applying different statistical methods.

9.4.1 Multivariate Linear Regression (MLR)

The most simple type of QSAR is a *Multivariate Linear Regression* (MLR):

$$\ln \mathrm{A} = \sum_i a_i \cdot p_i + \mathrm{B}$$

where A is the activity of the compound, p_i are the predictor variables, a_i the individual fitting parameters for the predictor variables, and B a fitting constant.

The activity is usually entered by its logarithm into such relationships, as this would prevent negative activities possibly resulting from other, new values of predictor variables for newly designed compounds. The fitting process is then performed by using a least-squares algorithm. If a number of possible predictor variables is available, one must test their individual statistical significance for the relationship and modify the model by systematically adding/removing them. In general, one should target a model with as few as possible predictors, and their number must also be limited with respect to the number of available test compounds.

Once a model with an acceptable *regression coefficient* R^2 has been identified, the aforementioned test for the *predictability* of the model must be performed. If either regression or predictability are poor, the MLR model is apparently too simple and nonlinear models must be employed.

9.4.2 Nonlinear Regression

Among the nonlinear regression models, *Alternate Conditional Expectations* (ACE), *Project Pursuit Regression* (PPR), and *Multivariate Adaptive Regression Splines* (MARS) are the most common. These can be characterised as follows.

9.4.2.1 Alternate Conditional Expectations (ACE)

The basic equation for ACE is

$$\Theta(Y) = \Phi_1(X_1) + \Phi_2(X_2) + \cdots + \Phi_p(X_p)$$

ACE provides nonlinear transformations on both the response and the predictors to minimize the mean squared error

$$e^2 = \frac{1}{n}\sum_{i=1}^{n}[\Theta(Y_i) - \sum_{j=1}^{p}\Phi_j(X_{ij})]^2$$

The ACE transformations are obtained using a two-dimensional scatterplot smoother. If $\Theta(Y) = Y$ and $\Phi_j(X_j) = \alpha_j X_j$ the ACE model becomes a linear model.

9.4.2.2 Project Pursuit Regression (PPR)

PPR is characterised by the equation

$$Y = \sum_{i=1}^{M}\beta_i f_i\left(\sum_{j=1}^{p}\alpha_{ij}X_j\right)$$

PPR can model the interaction between descriptors, as illustrated in Fig. 9.3. The corresponding mathematical formula is given by:

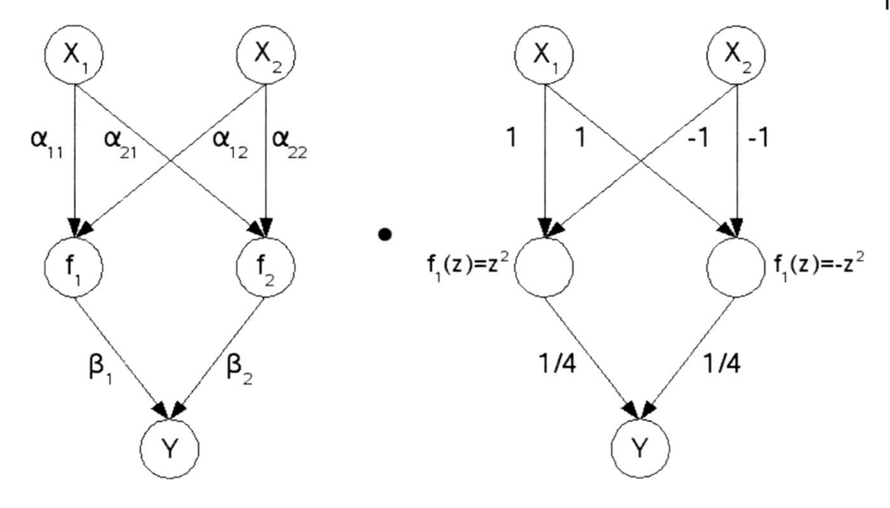

Fig. 9.3 Interaction scheme of parameters in the Project Pursuit Regression (PPR) method.

$$\begin{aligned} Y &= \beta_1 f(\alpha_{11}X_1 + \alpha_{12}X_2) + \beta_2 f(\alpha_{21}X_1 + \alpha_{22}X_2) \\ &= \frac{1}{4}(X_1 + X_2)^2 - \frac{1}{4}(X_1 - X_2)^2 \\ &= \frac{1}{2}X_1X_2 + \frac{1}{2}X_1X_2 \\ &= X_1X_2 \end{aligned}$$

If $M = 1$, $\beta_i = 1$ and $f_i(z) = z$, the PPR model will become a multiple linear regression model.

9.4.2.3 Multivariate Adaptive Regression Splines (MARS)

MARS is a local nonparametric regression method that builds up a set of tensor-product splines as basis functions, modelling the 'true' function $f(x)$ by:

$$\hat{f}(x) = a_\circ + \sum_{m=1}^{M} a_m \prod_{k=1}^{K_m} B_{km}(x_{v(k,m)}),$$

where $v(k, m)$ is the index of the factor used as argument of B_{km}, and K_m is a parameter that limits the order of interactions. The basis functions B_{km} are first-order truncated power splines defined by $B_{km}(x) = \pm(x - t_{km})_+$. t_{km} is a knot location chosen from the observed values of the corresponding components, and the function $(\ )_+$ delivers its value for positive arguments and sets it to zero otherwise. The key concept underlying the splines are the knots t, marking the end of one region of data and the beginning of another. In classical splines the knots are predetermined and evenly spaced, whereas in MARS the knots are determined by an iterative search procedure, which produces the basis functions leading to the smallest residual sum of squares in the regression. If all crossed terms were neglected and the basis functions replaced by a linear combination of the predictor variables, one would arrive at a simple multivariate linear regression again.

9.4.3 Example Calculation

Our example, which deals with the linear and two nonlinear regression methods, consists of a set of 27 known antibiotics of the cephalosporin type, the parent compound of which, cefpirome, is illustrated in Fig. 9.4.

Various substituents at the carbon atom number 10 in this formula modulate the antibacterial activity of this compound, and these differently substituted compounds have been made the basis of a QSAR study. For all numbered atoms the Mulliken populations have been extracted from *ab initio* calculations with a DZV basis set and used as potential predictor variables in the establishment of a relationship. The results are collected in Table 9.1. n indicates the number of descrip-

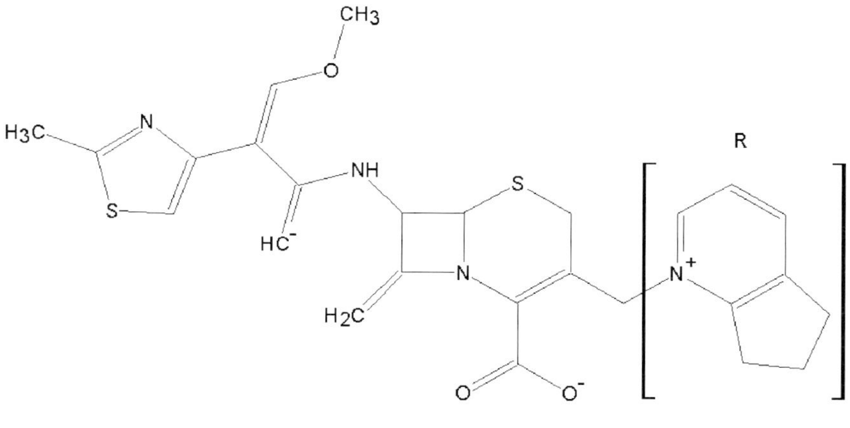

Fig. 9.4 Parent compound cefpirome for the QSAR example.

Table 9.1 Results of QSAR for cephalosporins based on quantum chemical data with different statistical correlation methods.

Method	*n*	Atoms	R^2	Q^2
MLR	7	S1, C3, C6, C7, O9, C10, N11	0.590	0.233
ACE	6	C2, C3, C7, C8, O9, C10	0.986	0.324
PPR	6	S1, C2, C3, C8, O9, N11	0.921	0.711

tors found significant to build a suitable model, namely the atoms listed in the table. R^2 is the correlation coefficient for the model, and Q^2 the normalised predictability of this model. It is clearly seen that it is impossible to construct a good linear model, as both fitting and predictability are highly unsatisfactory. Among the nonlinear models, ACE delivers an excellent fitting, but the predictability is much too low. PPR delivers an acceptable fitting and a predictability with a much higher confidence limit.

Interestingly, even the poorer models of MLR and ACE lead to almost the same descriptor variables as the PPR model, indicating where a change of electron density has the strongest influence on the activity of the drug, and thus pointing at the active centre of the drug.

This brief example provides only a limited insight into the broad field of QSAR. Many more sophisticated methods have been developed, for example, *Comparative Molecular Field Analysis* (COMFA), for which the reader is encouraged to consult specialised books on modelling and QSAR.

Another important component of more sophisticated modelling methods was mentioned earlier, namely statistical simulations using the Monte Carlo and/or the Molecular Dynamics methods. As these methods also provide many other important applications in computational chemistry, they are treated separately in Chapter 10.

Test Questions Related to this Chapter

1. Which are the most important contributions to force fields?
2. What are the principal procedures implemented in simulated annealing?
3. What could be the error sources in homology modelling?
4. How can we make use of quantum chemical results in predicting the properties of unknown molecules, and what are the limits of such a prodecure?
5. What statistical methods are usually employed to establish quantitative structure–property relationships?

10
Statistical Simulations: Monte Carlo and Molecular Dynamics Methods

Everything is in a flow, and there is no permanent reality except the reality of change.

Heraklitos, 6th century BC

10.1
Common Features

One general feature of chemical simulations is their foundation in classical physics and the treatment of large numbers of particles in order to achieve a significant sampling for their statistical evaluation. However, we will see that the purely classical treatment has its practical limits, and that quantum mechanics will enter the methodology whenever high accuracy and universality are required. These statistical methods deal with large ensembles of molecules in a dynamic equilibrium; several types of ensemble are described, with each abbreviated according to its statistical thermodynamics classification:

- NVE – constant number of particles, constant volume, constant energy (*microcanonical ensemble*).
- NVT – constant number of particles, constant volume, constant temperature (*canonical ensemble*).
- μVT – constant number of particles, constant chemical potential, constant temperature (*grand canonical ensemble*).

The second of these ensembles – NVT – is by far the most common for chemical simulations. Later, we will describe how this ensemble is managed in practice – that is, how, besides the number of particles and volume, the temperature is kept constant.

Statistical simulations are mainly employed for the treatment of liquid systems, where disorder of the gas phase combines with the density of solids, thus making the liquid state the most complicated for theoretical treatment. On the other hand, as most chemical reactions take place in solution, and biological processes

The Basics of Theoretical and Computational Chemistry. Edited by B. M. Rode, T. S. Hofer and M. D. Kugler

ISBN: 978-3-527-31773-8

occur almost exclusively in an aqueous environment, the liquid state is the most important one for chemistry, and a sufficiently accurate theoretical treatment of this state is, therefore, enormously important. Previously, while discussing force field modelling, we mentioned that the inclusion of solvent can be crucial for the quality of results, and the dynamic nature of solvation makes the statistical simulation methods – at least in a simple form – an indispensable instrument. However, in the case of pure liquids, electrolyte solutions or complexes in solution, the methodical requirements for a reliable description drastically increase, and in these situations simple models introducing averaged solvent effects [e.g., polarisable continuum models (PCM), implemented in some quantum chemical program packages], or describing solute–solvent interactions with relatively simplistic force fields (which still give satisfactory results for hydrated biomolecules) are bound to fail.

The initial steps in performing a statistical simulation are common to both the Monte Carlo (MC) and Molecular Dynamics (MD) methods, and include:

1. The choice of an appropriate 'elementary box'.
2. The definition of interaction potentials.
3. The generation of a starting structure.

These steps will be discussed before introducing the methodical differences between MC and MD.

The first of the steps is also inseparably connected with two other important features of the simulation of condensed liquid systems, the *periodic boundary condition* and the *minimal image convention*. The 'elementary box' – which usually is in the shape of a cube – must be an appropriate representation of the system to be studied, and contain a sufficient number of all involved species at an appropriate density. The latter is commonly derived or extrapolated from experimental data measured at the temperature for which the simulation is to be made. Figure 10.1 shows a typical example of a cubic elementary box for the simulation of liquid water at room temperature.

Even if this elementary box contains several hundreds or thousands of molecules, it would rather represent surface conditions than the situation in bulk. This is the reason for invoking the periodic boundary condition, which means that the elementary cube is assumed to be surrounded by identical cubes in all space directions (cf. Fig. 10.2), and that interactions between particles will take into account these adjacent cubes (and further cubes adjacent to them). For most types of interactions a cut-off can be applied, due to their rapid decrease with distance. However, this is not true for Coulombic interactions, which fall off with $\frac{1}{R}$; in order to fully consider these, special methods are implemented in the simulations, as we will see soon.

As the simulated system is characterised by the movements of particles, another important feature of the periodic boundary conditions is that whenever a particle leaves the elementary box, its image will enter from the opposite side, thus keeping the density constant. In order to ensure that no particle interacts

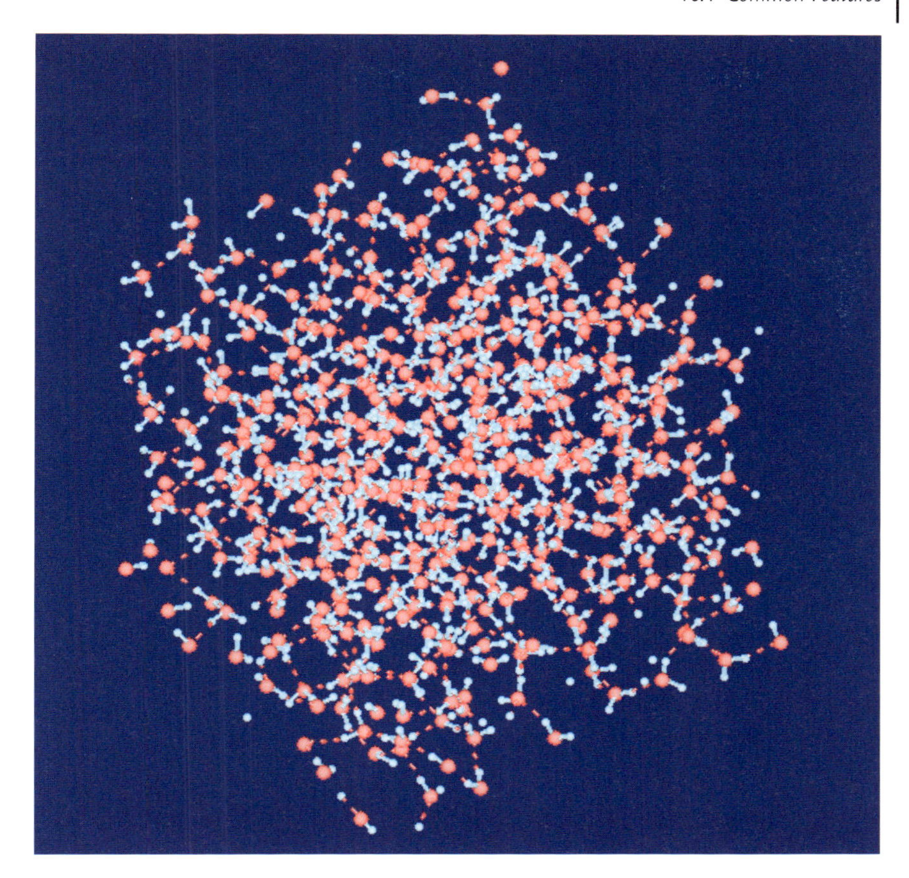

Fig. 10.1 Cubic elementary box for the simulation of liquid water.

with its images in the other boxes, and that interactions with other particles are always calculated with the nearest image of each of these particles, the minimal image convention is implemented in the simulation in order to consider the appropriate coordinates:

$$\text{if } r_{x,y,z}(i) \geq \frac{BL}{2} \Rightarrow r_{x,y,z}(i) = r_{x,y,z}(i) - BL$$

$$\text{if } r_{x,y,z}(i) \leq -\frac{BL}{2} \Rightarrow r_{x,y,z}(i) = r_{x,y,z}(i) + BL$$

where $r_{x,y,z}(i)$ are the Cartesian coordinates of the ith particle and BL is the length of the elementary box. This procedure is also illustrated in Fig. 10.2.

It is obvious that a too-small elementary box will impose an artificial symmetry effect on the simulated system. In practice, this means that for the description of

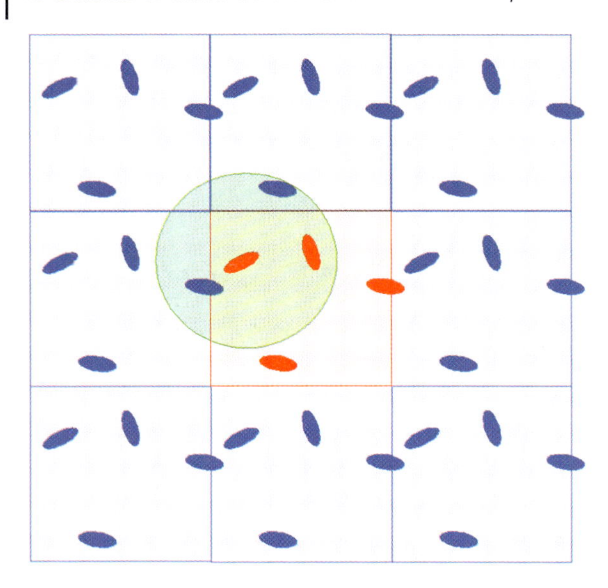

Fig. 10.2 Schematic representation of periodic boundary condition and minimal image convention.

short-range order, at least a few hundred particles must be contained in this box, but if one is interested in long-range order then this number must be extended accordingly.

When the system has been set up this way, one has then to determine how the interactions between particles in the box will be evaluated – that is, how to calculate the energy of the elementary box (MC) or energy and the forces acting on all particles in the elementary box (MD). This is usually achieved through interaction potentials which can either be obtained from empirical data or from *ab initio* calculations of interaction energy surfaces. A simple form of such a potential consists of a Coulombic term and a van der Waals term (R^{-6}, R^{-12}), but in practice these terms are often supplemented by further R^{-n} terms, with n being numbers between 4 and 11. These terms represent higher electrostatic interactions such as dipole–dipole, dipole–induced dipole, and similar interactions. Sometimes an exponential term which better accounts for certain repulsive interactions is also added. Interaction between molecules is usually considered on the basis of atom–atom interactions, summing over all atoms in each molecule. Thus, a very general form for such an analytical potential function describing pairwise interaction between two molecules is

$$\Delta E = \sum_{i}^{n} \sum_{j}^{m} \left[\frac{q_i q_j}{r_{ij}} + \frac{A_{ij}}{r_{ij}^a} + \frac{B_{ij}}{r_{ij}^b} + \frac{C_{ij}}{r_{ij}^c} + \frac{D_{ij}}{r_{ij}^d} + \cdots + X_{ij} e^{-Y_{ij} r_{ij}} \right]$$

where q_i represent (partial) charges, $A_{ij}, B_{ij} \ldots$ are fitting parameters and $a, b, c, \ldots$ exponents, usually in the range from 4 to 12. The summations run over all atoms i of the first and all atoms j of the second molecule. The parameters of these potential functions are fitted by a suitable algorithm (e.g., Levenberg–Marquart), to describe optimally the experimental or *ab initio*-calculated data.

One can see that most of these terms decay rapidly and would become $\sim\emptyset$ within the elementary box. For some of them an even shorter cut-off than half of the box length can be applied. This is not the case, however, for the Coulombic terms, which remain relevant beyond the limits of the elementary box. When considering these *long-range contributions*, two principal methods are available: (i) the Ewald summation; and (ii) the Reaction Field method:

- The *Ewald summation* evaluates all electrostatic interactions of particles inside the box with those in the periodically repeating boxes, and a formalism making use of the reciprocal lattice provides a relatively rapid convergence. This formalism is usually only applicable to electroneutral systems, but newer modifications make it possible to overcome this restriction.
- The *Reaction Field method* evaluates a 'field' based on the (partial) charges of all particles in the outside boxes and, taking into account the dielectric constant of the liquid medium, evaluates the long-range interactions of all particles inside the box with those outside as an average contribution.

The pairwise interaction between particles is only the first term in a series of terms needed for an accurate description of interactions in a dense, multi-particle system:

$$\Delta E = \sum_{a,b} \Delta e^{pair} + \sum_{a,b,c} \Delta e^{3\text{-}body} + \sum_{a,b,c,d} \Delta e^{4\text{-}body} + \cdots + \sum_{a,b,c,d,\ldots n} \Delta e^{n\text{-}body}$$

The stronger the interactions – and thus the mutual polarisation effects – are, the more relevant the higher terms (i.e., 3-body, 4-body or, generally n-body) become. It is already quite difficult to find suitable analytical potential functions that reflect these higher terms for 3-body corrections, and in most cases it is almost impossible for higher n-body terms. For example, the *ab initio* construction of a 3-body correction function requires the calculation of around 10 000 points of the corresponding interaction energy surface, and it can prove quite difficult to find and fit a suitable analytical function that accurately reflects this energy surface. There are other – albeit much more computer intensive – ways to include higher interaction terms, which have become accessible to a larger extent only recently; these will be discussed later in this chapter.

The final step in the preparation of these simulations is the choice of a starting structure, i.e. the distribution of particles within the elementary box. The most general approach is to start from a randomly generated distribution, but this can be quite time-consuming as such a configuration is far from equilibrium and many steps will be needed to reach an equilibrated configuration. If a solid-state configuration is known, this will be a better starting solution; moreover, if an equilibrated similar liquid system is available, this will be even more suitable to create an initial configuration.

We have now reached the point where the MC and MD methods diverge, and similarities will only be encountered again when sampling is finished and data evaluation begins.

10.2 Monte Carlo Simulations

Monte Carlo simulations are based on the production of a large number of randomly generated configurations of the system and their subsequent statistical evaluation. The generated configurations are linked together in a so-called *Markov chain*; that is, every new configuration is evolving from the previous one. In practice, a randomly selected particle in the elementary box is subjected to a random translation and rotation, and the energy of the new configuration is then compared to the energy of the previous configuration. If the new energy is lower, the configuration is accepted, but if it is higher then the Boltzmann factor $f_B = e^{-\frac{\Delta E}{kT}}$ is calculated, where ΔE is the energy difference between new and old configuration. f_B is then compared with a random number r between 0 and 1, and if f_B is larger than r, the new configuration is accepted; otherwise, it is rejected. This procedure is known as the *Metropolis algorithm*, and ensures that a certain amount of energetically less-favourable configurations will also be taken into account, corresponding to the changes occurring in a real liquid. By using this algorithm one has also introduced (constant) temperature into the system, and it is evident that the higher the temperature, the more configurations with a higher energy will be accepted. By adjusting some parameters of the simulation – mainly the step size for translations and rotations – one can (automatically) achieve a general acceptance/rejection rate, which is ideally around 1/3 in order to obtain a representative sampling.

The energy of the total system will initially decrease quite rapidly, if one has started from a configuration far from equilibrium. At the end of the equilibration process, the energy will begin to fluctuate around an average value (Fig. 10.3).

In order to ensure that this is a real equilibrium state, it is recommended that the temperature is increased for 10 000 to 30 000 steps to a few thousand degrees K (this will bring the system to a much more disordered state), and is then reset to the desired simulation temperature. If the average value of energy is the same as before, it can be assumed that the system is in a real equilibrium (cf. Fig. 10.3). The total equilibration procedure including this test can require the generation of a few hundred thousand configurations, and sampling of the system only starts after this has been accomplished. For this sampling, usually every 500th to

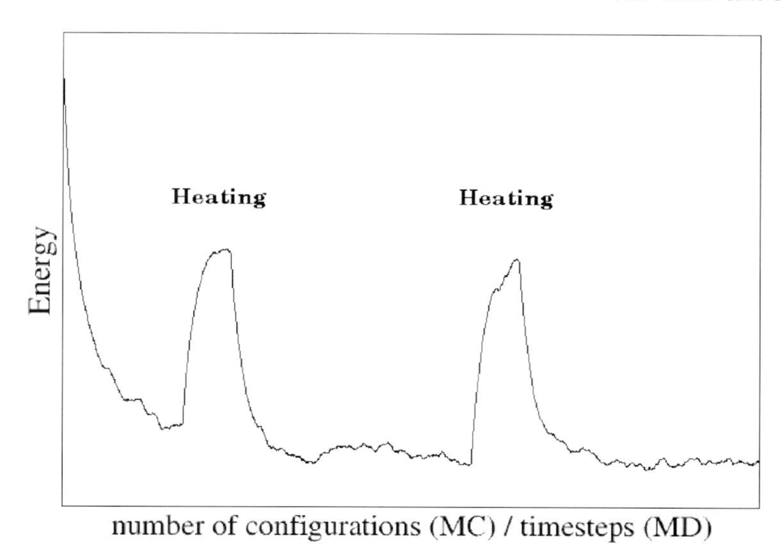

Fig. 10.3 Energy evolution during a simulation, showing the equilibration process. Two short intermediate heating processes have been performed to assure the establishment of a 'true' equlibrium configuration.

1000th out of a few millions random-generated configurations is taken and stored in the so-called *'history file'* of the simulation, which contains the coordinates of all atoms and the associated energies for the whole system. This history file forms the basis for all further data evaluations.

In Chapter 9 it was mentioned that Monte Carlo methods play an important role when the interaction of molecules is modelled, particularly in the procedure of simulated annealing. The Monte Carlo method has the great advantage that temperature can be changed simply by an input variable (we will see that temperature control is a more complicated process in MD), and the amount of displacement and rotation of the molecules allowed in one step can be adjusted just as simply via input parameters. This allows a stepwise 'scanning' of the interaction with a gradually finer grid. However, the size of the macromolecules imposes a strong restriction on the method, as relatively simple empirical force field potentials must be used to describe the molecular interactions. As mentioned above, a simulation of a macromolecule and its interaction with other molecules in a biological environment must incorporate a sufficient quantity of water molecules and, eventually, also counterions present in the surrounding of the biopolymer. For interaction with these species it is also necessary to use more or less empirical and simple potentials, thus restricting the accuracy of such simulations.

In the case of molecular liquids or solutions, the potentials to be employed can be considerably more sophisticated; in addition to the terms given above as the general form of pair potentials, they can contain 3-body correction functions or other terms taking into account the polarisability. In the latter case one also speaks of '*polarisable potential functions*', and the search for suitable potentials of this type has occupied many research groups during the past two decades. Nowadays, quantum mechanics is increasingly entering simulation protocols, thus making classical n-body and polarisation corrections slowly but steadily less important. Such quantum mechanical simulations are detailed later in this chapter.

10.3 Molecular Dynamics Simulations

Molecular dynamics (MD) consider an ensemble of particles and all forces acting on these particles, attempting to solve the equations of motion for all the particles, and targeting a unique trajectory predicting all future (and past) states of the system (at least for a classical system without the uncertainties of quantum effects). The main difference in this theoretical approach compared to Monte Carlo methods is the inclusion of the time variable, which enables the calculation of much more data for the system, namely those related to time and the system dynamics.

On the basis of the Lagrangian function $L = T - V$ (i.e., the difference between kinetic and potential energy), the Lagrangian equation of motion is defined by

$$\frac{d}{dt}\left(\frac{\partial L}{\partial \frac{\partial q_k}{\partial t}}\right) - \left(\frac{\partial L}{\partial q_k}\right) = \emptyset$$

where q_k are the generalised coordinates of the particles. Considering an ensemble of atoms defined in Cartesian coordinates with the vectors $|r_i\rangle$ we obtain the expression

$$m_i \frac{\partial^2 |r_i\rangle}{\partial t^2} = |f_i\rangle$$

where m_i is the mass of the i_{th} atom and $|f_i\rangle$ specified as

$$|f_i\rangle = \nabla_{r_i} L = -\nabla_{r_i} V$$

is the force acting on this atom. The same formalism can be applied to molecules, where the centre of mass represents its 'coordinate' and $|f_i\rangle$ the total force acting on this molecule.

In order to compute trajectories for N-particle systems, one would have to solve 3 N second-order differential equations. In practical MD simulations, however, this task is replaced by the *finite difference approach*, by which we try to determine from the information (positions of particles, velocities) on the system at a given time t the values of these quantities at a time $t + \Delta t$. Δt must be significantly smaller than the time of a typical movement of a particle in the system (e.g., a molecular vibration or a translational movement of a molecule corresponding to the size of this molecule). For the practical implementation of this procedure the two most frequently used methods will be given as examples – the so-called *predictor–corrector algorithm* and the *Verlet algorithm*.

For a continuous trajectory, the estimated values for positions, velocities, and other properties such as accelerations a and third-time derivatives of $|r_i\rangle$ b can be obtained by a Taylor expansion about time:

$$\begin{aligned}
|r^p(t+\delta t)\rangle &= |r(t)\rangle + \delta t|v(t)\rangle + \tfrac{1}{2}\delta t^2|a(t)\rangle + \tfrac{1}{6}\delta t^3|b(t)\rangle + \cdots \\
|v^p(t+\delta t)\rangle &= |v(t)\rangle + \delta t|a(t)\rangle + \tfrac{1}{2}\delta t^2|b(t)\rangle + \cdots \\
|a^p(t+\delta t)\rangle &= |a(t)\rangle + \delta t|b(t)\rangle + \cdots \\
|b^p(t+\delta t)\rangle &= |b(t)\rangle + \cdots
\end{aligned}$$

Here, the superscript p classifies these values as 'predicted' values. We truncate the expansions as shown above and store the vectors $|r\rangle$, $|v\rangle$, $|a\rangle$ and $|b\rangle$. However, this formalism will not give really correct trajectories, as the equations of motion have not been introduced. Therefore, a second step – the 'corrector' step – has to be implemented. From the new positions of the particles $|r^p\rangle$ we calculate the forces at time $t+\Delta t$ and hence the correct accelerations $a^c(t+\Delta t)$, which can be compared with the predicted accelerations $a^p(t+\Delta t)$ to estimate the error in the prediction step, which is then implemented in the corrector step:

$$\begin{aligned}
\Delta|a(t+\delta t)\rangle &= |a^c(t+\delta t)\rangle - |a^p(t+\delta t)\rangle \\
|r^c(t+\delta t)\rangle &= |r^p(t+\delta t)\rangle + c_0\Delta|a(t+\delta t)\rangle \\
|v^c(t+\delta t)\rangle &= |v^p(t+\delta t)\rangle + c_1\Delta|a(t+\delta t)\rangle \\
|a^c(t+\delta t)\rangle &= |a^p(t+\delta t)\rangle + c_2\Delta|a(t+\delta t)\rangle \\
|b^c(t+\delta t)\rangle &= |b^p(t+\delta t)\rangle + c_3\Delta|a(t+\delta t)\rangle
\end{aligned}$$

The coefficients c_i have been optimised for different orders of differential equations, and the numbers of position derivatives needed can be found in the literature for given conditions.

Thus, the predictor–corrector algorithm is applied in the following steps:

(a) predict the positions, velocities, accelerations, etc., at a time $t+\delta t$, using the current and previous values of these quantities;
(b) evaluate the forces, and hence accelerations $|a_i\rangle = \frac{1}{m_i}|f_i\rangle$, from the new positions;
(c) correct the predicted positions, velocities, etc., using the new accelerations; and
(d) calculate any variables of interest for the accumulation of time averages, before returning to (a) for the next step.

The Verlet algorithm is a direct solution of the equation

$$m_i \frac{\partial^2 |r_i\rangle}{\partial t^2} = |f_i\rangle$$

based on the positions $|r_t\rangle$, accelerations $|a_t\rangle$ and the positions $|r_{t-\Delta t}\rangle$ of the previous step. The equations to obtain the advanced positions are given as

$$\begin{aligned}
|r(t+\delta t)\rangle &= 2|r(t)\rangle - |r(t-\delta t)\rangle + \delta t^2 |a(t)\rangle \\
|r(t+\delta t)\rangle &= |r(t)\rangle + \delta t|v(t)\rangle + \tfrac{1}{2}\delta t^2 |a(t)\rangle + \cdots \\
|r(t-\delta t)\rangle &= |r(t)\rangle - \delta t|v(t)\rangle + \tfrac{1}{2}\delta t^2 |a(t)\rangle - \cdots
\end{aligned}$$

in which the velocities have been eliminated by addition of the equations obtained by the Taylor expansion about $|r_t\rangle$.

Although the velocities are not needed to determine the trajectory, they are required to extract important properties of the simulated system from the simulation, and hence will always be calculated by

$$|v(t)\rangle = \frac{|r(t+\delta t)\rangle - |r(t-\delta t)\rangle}{2\delta t}$$

The *leap-frog algorithm* is a modification of the Verlet algorithm employing half-step widths $\frac{\Delta t}{2}$.

For more details and practical advantages/disadvantages of these algorithms, the reader should refer to specialised books. At this point it should be mentioned, however, that the time step Δt is a crucial choice, as it determines the accuracy of the data to be evaluated by the simulation. For example, if explicit hydrogen movements in a molecule are to be reproduced correctly, a time step of 0.2 femtoseconds is appropriate, whereas for the movements of the backbone of a protein the step may be up to a few femtoseconds. By considering the over-all computational effort, this will determine the total time span that can be covered by a simulation.

The above-discussed algorithms impose a condition on the intermolecular potential functions describing the interactions between the particles, be it pair or n-body potentials: they must be differentiable, and one must keep this condition in mind when developing analytical functions to describe the energy surfaces.

When starting the MD simulation from an initial configuration, a sufficiently large number of time steps must be provided for achieving equilibration – typically a few ten to hundred thousands. There are some important differences between the equilibration procedure of MC and MD simulations. Whilst in an MC simulation *one* particle is moved/rotated in every step, in an MD simulation *all* particles of the system move simultaneously. A second important difference is the way that temperature is handled. In the MC simulation, the Boltzmann factor in the Metropolis algorithm controls the temperature, but in the

MD simulation there is no such control factor to ensure a constant temperature for the NVT ensemble. According to statistical thermodynamics, temperature is a function of the velocities of the particles, and this is the easiest way to control temperature, namely by scaling the velocities of all particles to keep temperature constant. This process – known as the *Berendsen* algorithm – corresponds to the action of an external bath to guarantee isothermal conditions. This scaling is applied in every time step, starting from the evaluation of the kinetic energy:

$$E_{kin} = \sum_{i}^{N} \frac{m_i v_i^2}{2} = \frac{3}{2} NKT$$

which delivers the actual temperature T. The scaling factor for the velocities λ is then obtained as

$$\lambda = \sqrt{1 + \frac{\Delta t}{\tau}\left(\frac{T}{T_\circ} - 1\right)}$$

where τ is a relaxation time, usually chosen in the subpico- or pico-second range, and $T_\circ$ the temperature to be maintained.

Although other, equivalent procedures are available to control temperature, the Berendsen algorithm seems to be the most commonly used.

When equilibration has been accomplished, the sampling phase starts. The number of time steps needed for sampling depends on the number of different species and their percentual occurrence in the system, as well as on the type of data one wants to extract from the simulation. In particular, if solution dynamics or structural behaviour of macromolecules in time are the target of the investigation, then long simulation times become unavoidable.

In the MD simulations the *trajectory* is equivalent to the 'history file' of MC simulations in which, besides coordinates of all atoms and energies, the velocities of the particles and other dynamic data are also stored, thus significantly increasing the total amount of storage media needed. On the other hand, these additional data enable us to obtain much more information about the simulated system than any MC simulation could provide. This refers in particular to all time-dependent dynamic data, and we will see the additional power of MD simulations in detail in the following section.

Molecular dynamics of macromolecules rely basically on the same force fields as their Monte Carlo analogues. Their advantage is the possibility of observing a time evolution of the system studied (which requires, however, the calculation of a very large number of time steps!), and their problem is to obtain equilibrium structures within a reasonable time (or at all), if the starting structure was a bad guess. Also, in the case of MD simulations of biologically relevant systems, the incorporation of water and – where applicable – counterions is essential for a good result. Figure 10.4 shows the results of MD simulations of the oligopeptide heptaglycine *in vacuo* (above) and in water (below), after the same number of

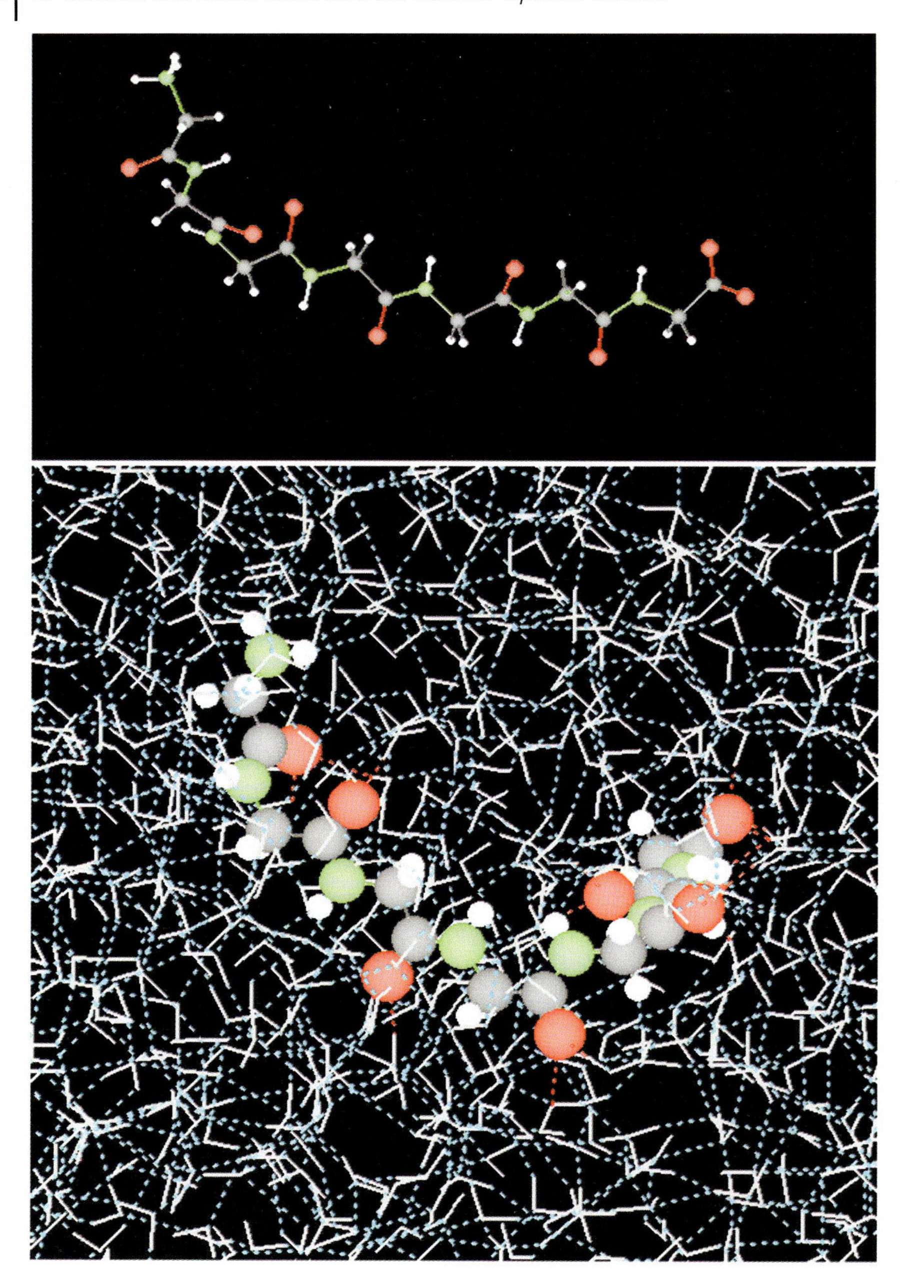

Fig. 10.4 MD simulation of the oligopeptide heptaglycine *in vacuo* (upper) and in water (lower), after the same number of time steps.

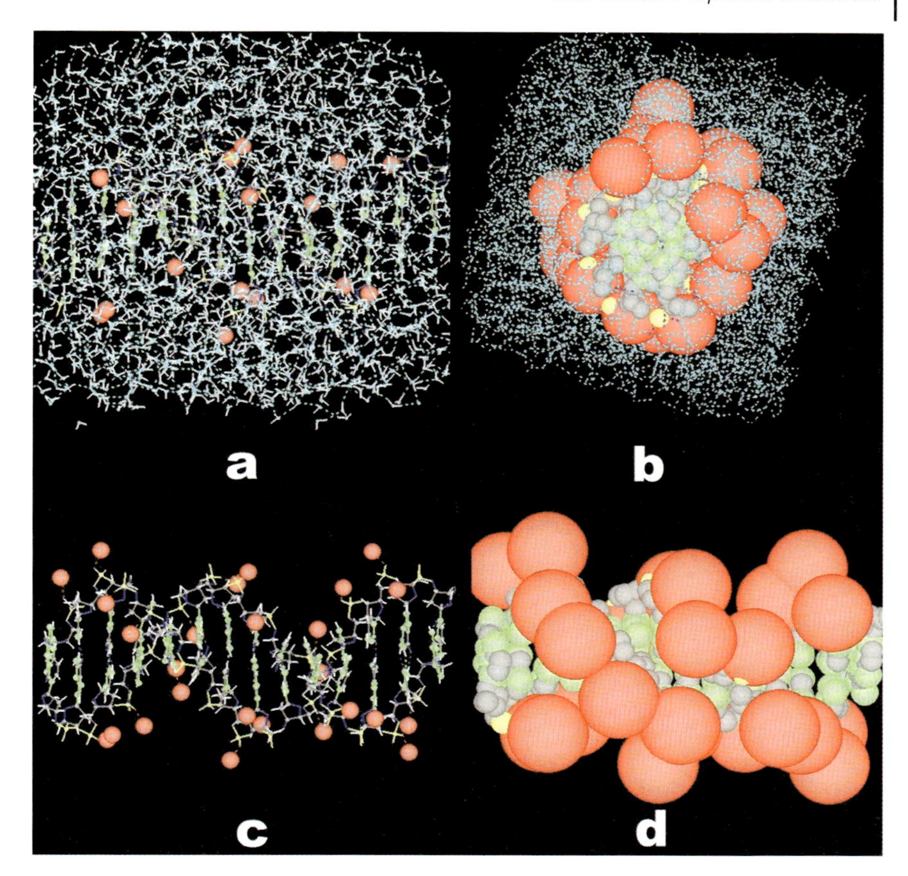

Fig. 10.5 Snapshot pictures from a simulation of a DNA fragment in water with Na^+ counterions (red atoms). Panels (a) and (b) show the DNA with the surrounding water; in panels (c) and (d) the solvent molecules have been made invisible. In (b) and (d) the Na^+ ions have been enlarged to represent the radius, up to which the $Na^+–H_2O$ binding force is stronger than any hydrogen bond.

time steps and with the same simulation protocol. Clearly, evolution of the peptide structure is considerably different under the influence of the solvent.

The importance of counterions is illustrated in Fig. 10.5, which shows snapshot pictures from a simulation of a DNA fragment in water with Na^+ counterions (red atoms). Figure 10.5(a) and (b) show the DNA with the surrounding water; in (c) and (d) the solvent molecules have been made invisible. In (b) and (d) the Na^+ ions have been enlarged to represent the radius, up to which the $Na^+–H_2O$ binding force is stronger than any hydrogen bond. These pictures demonstrate

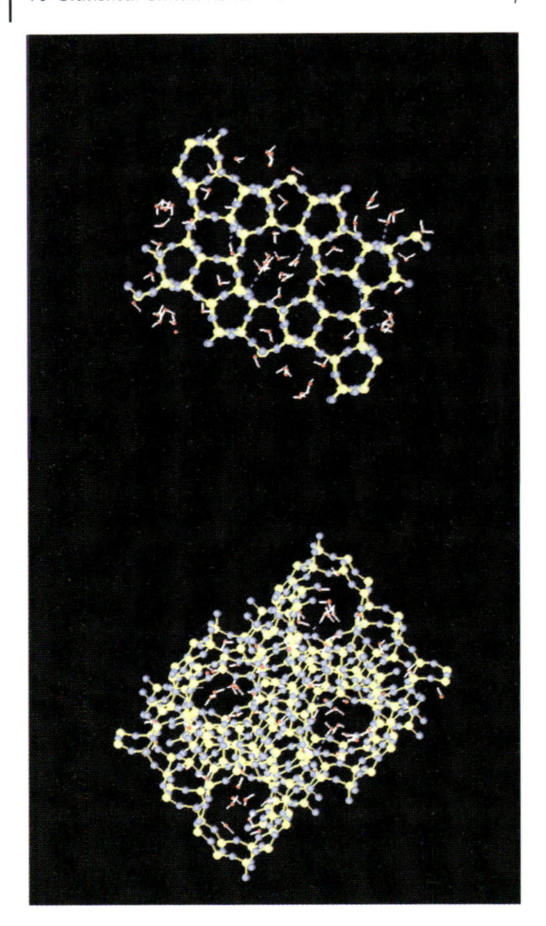

Fig. 10.6 Simulation of the distribution of water molecules inside a silica crystal.

that the gradual interaction of DNA with any reaction partner cannot be described without taking into account both the solvent and the ions.

Another example of simulation is depicted in Fig. 10.6, in which case the distribution of water inside a silica crystal has been investigated. The rigid solid remains in a fixed position throughout the simulation, and only the water molecules are allowed to move, revealing the dynamics of the solvent inside the channels formed by the crystal lattice. This example shows that simulations are also well suited for the simulation of interfaces such as solid/liquid or solid/gas.

As mentioned above in connection with modelling procedures and MC simulations, quantum mechanics applied to at least a part of the system now represent a significant contribution to the improvement of accuracy, and the formalism employed in such simulations will be discussed below.

10.4 Evaluation and Visualisation of Simulation Results

In this section, we will describe which physically and chemically relevant information can be obtained from the history files (MC) and trajectories (MD) produced by the simulations. Section 10.4.1 relates to structural data, which are equally available from both types of simulation, whereas the time-related data discussed in Section 10.4.2 are available only from MD simulations. In both parts the possibilities of visualising the results will be demonstrated, and experimental methods used to compare simulation results with spectroscopic measurements discussed.

10.4.1 Structure

Structural entities in a dynamically changing system, be it in the gaseous or liquid state, can be recognised from particle distributions deviating from a purely statistical distribution. Therefore, distribution functions provide the means to recognise such structural entities and to analyse them in detail. In the case of molecular particles, one can either use the centre of mass of the molecules to describe their distribution or – to enable a much more accurate analysis – the pair distribution functions between all atoms in these molecules and the atoms of other particles. This description by atom–atom pair distributions supplies information not only about the structural entities formed but also about the orientation of molecules in these entities and the binding processes responsible for structural order in the system.

The most important pair distributions are the *radial distribution functions (RDF)*, where the deviations from purely statistical distributions are plotted as a function of the distance between two particles or two atom types, respectively. Figure 10.7 shows the ion–O and ion–H radial distribution functions for the metal cation Al^{3+} in aqueous solution.

One can immediately recognise the formation of two distinct hydration shells and a less-pronounced third shell around the ion, and by comparing the maxima of ion–O RDF and ion–H RDF one can clearly derive the orientation of the water molecules in the first shell as dipole-oriented, with the O-atoms coordinated to the metal ion. The integration of the RDFs delivers the number of atoms (and hence the ligands) corresponding to the peaks, and thus the average coordination number of the ion–water complex.

More information can be extracted via *angular distribution functions (ADF)* for various angles (e.g., the angle between different ion–O vectors). The corresponding ADF for the first hydration shell of our example is shown in Fig. 10.8.

The peak maxima show the preferred orientation(s) of the ligands, and the peak width is a measure of the flexibility of ligand orientation. Another ADF could be employed to determine the rotational flexibility of the ligands, and in this way the combination of RDFs and ADFs will describe all structural features

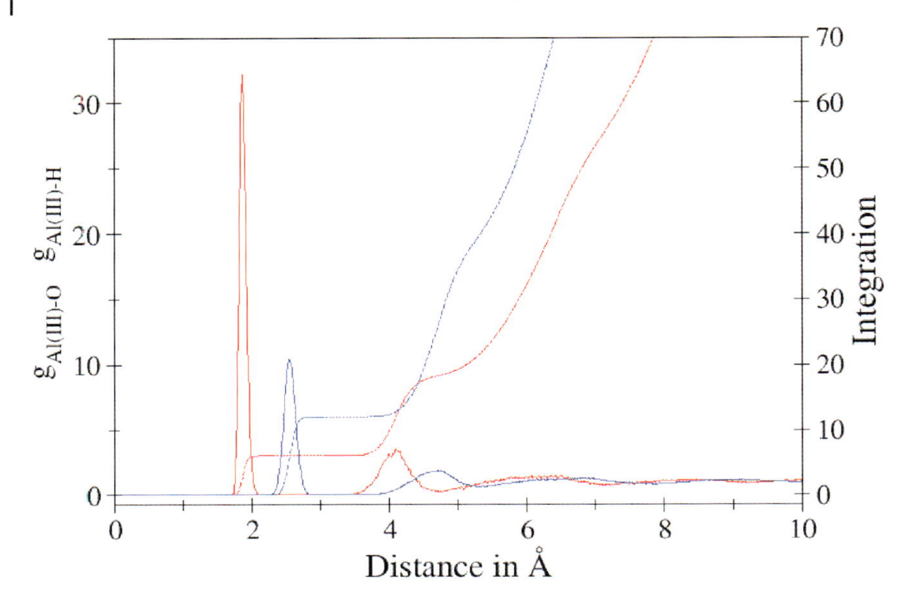

Fig. 10.7 Ion–O (red) and ion–H (blue) radial distribution functions and corresponding integration for Al^{3+} cation in aqueous solution.

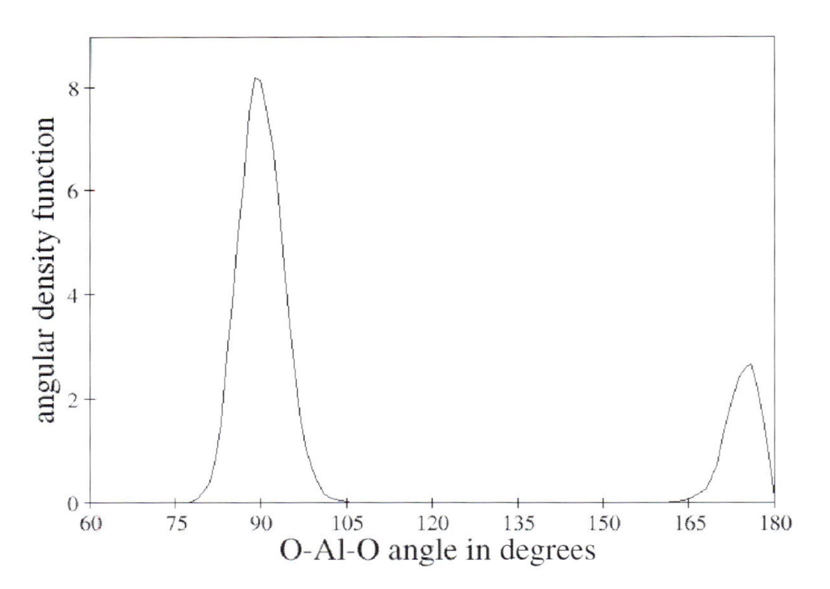

Fig. 10.8 Angular distribution function of ion–ligand orientation in the first hycration shell of Al^{3+} cation in aqueous solution.

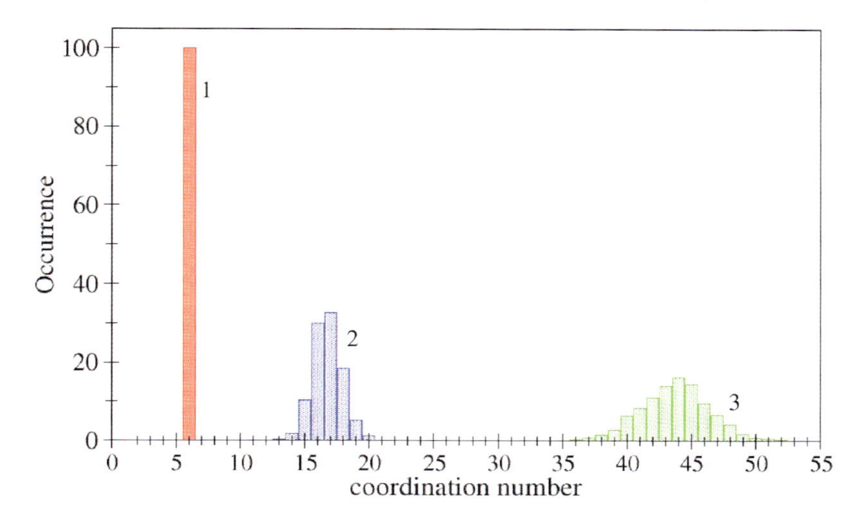

Fig. 10.9 Coordination number distributions of first (1), second (2) and third (3) hydration shell of Al^{3+} ion in aqueous solution.

in every detail needed to characterise even complicated multi-species systems in an unambiguous and complete manner.

Many systems are changing their structural features dynamically, often within such a short time span, that one must consider a number of structural entities to be simultaneously present for any kind of experimental measurement carried out on such systems. This can be illustrated by the evaluation of the *coordination number distribution (CND)* in our example, depicted in Fig. 10.9.

While the composition of the first hydration shell is stable within the whole simulation time ($N = 6$), the composition of the second and third shells is by no means constant, displaying a number of species with different numbers of ligands. The RDF delivers only average coordination numbers, and hence the detailed evaluation of history file or trajectory producing CNDs is an important supplementary tool to determine species distributions in the system under investigation. A further step to elucidate structural details is the specific evaluation of RDFs and ADFs for the differently coordinated species identified via the CND, which will allow the separate determination of bond distances and angular orientations for all species present. This is of particular importance if the first coordination shell is very labile, exchanging ligands on the picosecond scale.

However, it is evident that the quality of these evaluations depends on the correctness and completeness of the potentials employed in the simulation. There are a number of experimental data which can serve as control parameters for the quality of a simulation. The most accurate data are supplied by diffraction methods. X-ray diffraction (XRD) and neutron diffraction (ND) require relatively high concentrations, and thus do not always correspond to the conditions of simula-

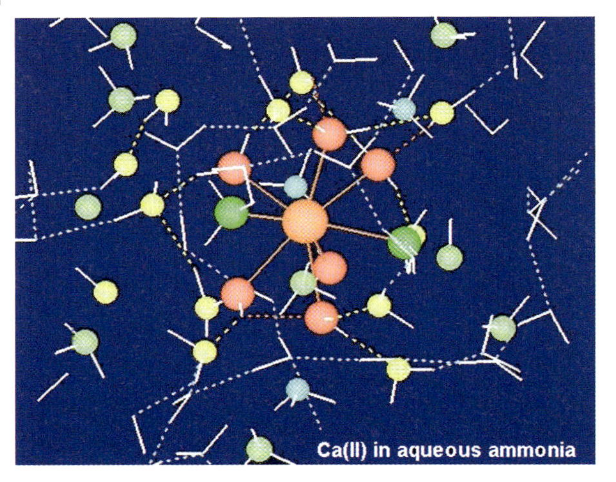

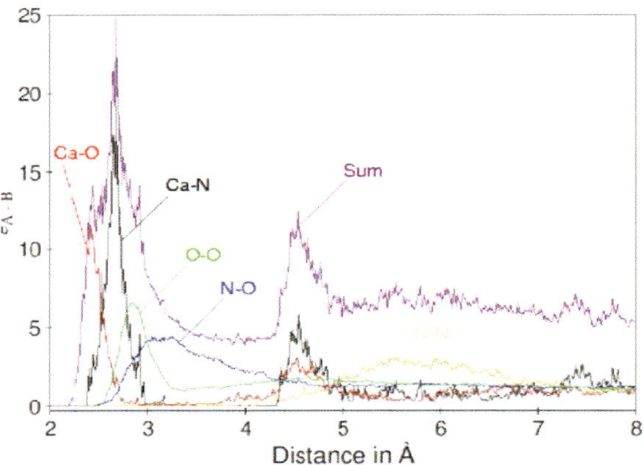

Fig. 10.10 Ca^{2+} ion in aqueous ammonia: snapshot picture from MD simulation and various radial distribution functions (RDF) including the sum of all RDFs.

tions. Of the two methods, ND is the more universal, as it will also show the positions of the hydrogen atoms. It also allows a more detailed analysis by the use of isotope substitution, which facilitates the assignment of observed spectral lines to specific atom pairs. However, this combined technique (NDIS) makes the experimental setup much more complicated (and expensive).

More recently, newer spectroscopic techniques such as Extended X-ray Absorption Fine Structure (EXAFS) and Large Angle X-ray Scattering (LAXS), which can function with much more dilute solutions, have strongly facilitated the comparison of simulation results with experiment. It should not be forgetten, however, that all of these techniques are based on fitting procedures, in which different models for the system are investigated for their compatibility with the measured spectra consisting of peaks for all atom–atom distances occurring in a dynamically changing scenario. The more complex the system becomes and the faster the interchange between species occurs, the more difficult it becomes to guess what would be a good model for the fitting procedure. This can be easily illustrated by the still relatively simple example shown in Fig. 10.10, taken from a simulation of a Ca^{2+} ion in aqueous ammonia solution.

Besides a snapshot picture of the solvated ion, Fig. 10.10 shows the various atom–atom RDFs, which overlap in most of the crucial regions where the structural data have to be extracted from any kind of diffraction pattern. The sum of all RDFs, representing the pattern visible to the spectroscopist, is also depicted, illustrating the difficulties encountered in an experimental investigation. In particular, the overlap of Ca–O, Ca–N and O–O RDFs makes it extremely difficult – without obtaining good advice from a simulation! – to construct appropriate models to fit the measured data. For the second solvation shell, any attempt to do so seems almost hopeless. In addition, the rapid ligand exchange at the cation leading to the presence of several different species in the solution, further complicates interpretation of the experimental data. This example, and others that will be given when discussing ultrafast dynamics of liquid systems, show that computational chemistry methods have not only become indispensable instruments supplementing experiment, but can sometimes even be superior to any experimental method – provided that the necessary methodical accuracy of the theoretical approach is achieved.

10.4.2 Dynamics

While the structural data can be obtained by either MC or MD simulations, it is clear that only MD can provide access to time-dependent data needed for an evaluation of the dynamics of a chemical system. Even in MD simulations, there are limits set by the feasible length of the trajectory: the faster the dynamics, the more suitable are the processes for simulation work and the more accurately the investigation can be performed; typically, reactions occurring in the pico- to millisecond range are the domain of MD simulations. Fortunately, a large percentage of the chemically and biologically relevant reactions take place within this time scale, and are thus accessible to theoretical investigations based on such simulations.

Besides simple dynamic quantities such as diffusion phenomena, the access to *rovibrational spectra* is a major advantage of MD simulations. Vibrational spectra are obtained via the *velocity autocorrelation functions (VACF)* and normal coordinate analysis. The normalised VACF is defined as

$$C(t) = \frac{\sum_i^{N_t} \sum_j^N v_j(t_i) v_j(t_i + t)}{N_t N \sum_i^{N_t} \sum_j^N v_j(t_i) v_j(t_i)},$$

where N is the number of particles, N_t is the number of time origins t_i, and v_j

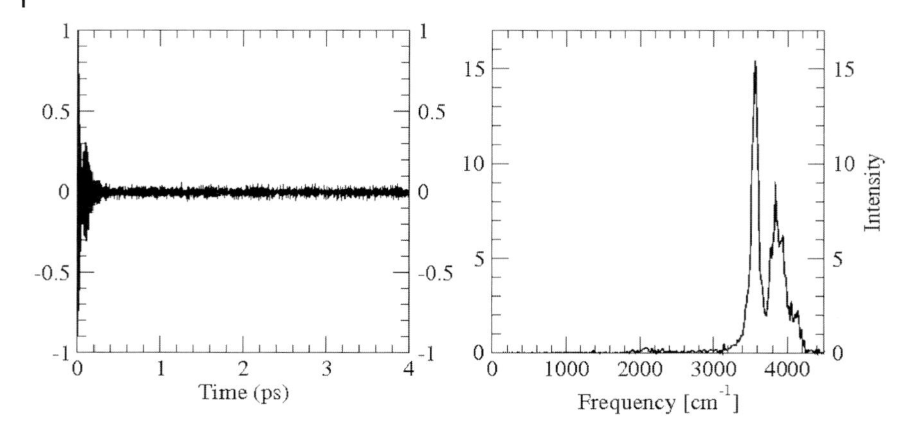

Fig. 10.11 Velocity autocorrelation function and associated vibrational spectrum of the symmetric stretching vibration of liquid water obtained by MD simulation.

denotes a certain velocity component of the particle j. A correlation length of 2.0 ps is typical to obtain the power spectra with a few thousand averaged time origins. Fourier transformation of the VACF delivers the vibrational spectrum. Figure 10.11 shows an example, depicting the VACF and associated vibrational spectrum of the symmetric stretching vibration of liquid water. The simulation also offers the advantage of specifying a selected region of the system for the evaluation of the spectra (e.g., the immediate surrounding of an active centre or the first coordination shell of a metal ion), in order to study the influence of the center on the ligands/solvent molecules directly bound to it.

MD simulations also allow the investigation of *reorientational time correlation functions (RTCF)* defined as

$$C_{li}(t) = \langle P_l(\vec{u}_i(0)\vec{u}_i(t))\rangle,$$

where P_l is the Legendre polynomial of lth order and $\vec{u}_i$ is a unit vector along three principal axes i defined in a fixed coordinate frame. RTCFs can then be fitted to the form

$$C_l(t) = a \cdot exp(-t/\tau_l),$$

where a and τ_l are the fitting parameters, and τ_l corresponds to the relaxation time. The correlation functions for $l = 1$ are related to IR lines, for $l = 2$ to Raman line shapes and NMR relaxation. As most of these parameters can be determined experimentally, they offer criteria to evaluate the quality of the simulation (or, sometimes, of the experimental work).

One of the most prominent topics in connection with MD simulations is that of reaction dynamics. Reaction rates and mechanisms can be studied with the help

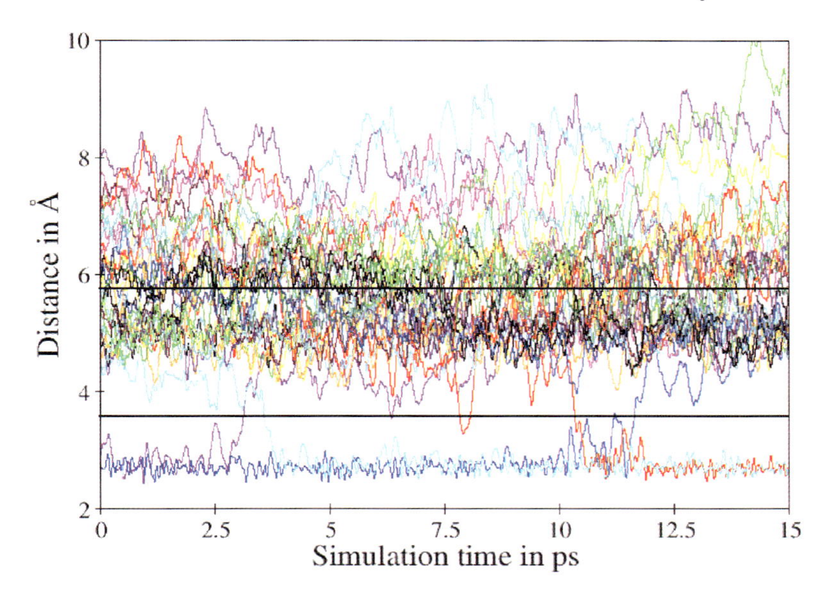

Fig. 10.12 Ion–ligand distance plot from a 15 ps trajectory for Sr^{2+} ion in water, showing the exchange processes between hydration shells and bulk solvent.

of trajectories, if the time scale of the reactions coincides with that of the simulation. This will again be demonstrated by a simple example, namely the ligand exchange reactions of a solvated ion. Figure 10.12 displays a diagram of all ion–ligand distances during the course of a 10 ps trajectory for Sr^{2+} ion in water. From the number of exchange events one can determine the mean residence time (MRT) of a ligand in the hydration shells, and from the way in which the outgoing and incoming ligands behave, one can classify the exchange mechanism according to the common scheme as associative, dissociative, or interchange type. In our example (Fig. 10.12), ligand exchanges in the first and second hydration shells of the Sr^{2+} ion are shown. An interesting aspect of this example is the subsequent occurrence of two different ligand exchange mechanisms. The first exchange is a typical dissociative one, while the second one is a 'classical' associative exchange. With present experimental means, this behaviour could never be revealed. The vigorous exchanges of ligands between second shell and bulk are clearly recognised.

As this hydrate shows, reaction dynamics can occur on an extremely short time scale. In such cases, comparison with experimental data is still limited to a few systems, where the new technique of femtosecond laser pulse spectroscopy is feasible, as the NMR time scale is limited to $\sim 10^{-9}$ seconds, and quasielastic neutron scattering reaches only $\sim 10^{-10}$ seconds. For comparison, the lifetime of hydrogen bonds, which governs many chemical systems, is in the sub-picosecond range (in water $\sim 5 \cdot 10^{-13} s$). Theoretical simulations are, therefore, sometimes the only method providing access to the ultrafast dynamics in liquid systems. In such cases, the accuracy of the intermolecular potentials is of particular importance for obtaining reliable data, and a careful assessment of the simulation protocol and its validity for the investigated system is mandatory. The need for such highly accurate simulation methods leads us again to the aforementioned incorporation of quantum mechanics into the statistical simulations.

While discussing the evaluation of simulation data, several types of graphical presentation have proven essential for the visualisation of results. In order to obtain a realistic impression of the dynamics of a system, still pictures (i.e., snapshots) can be helpful, as they illustrate various species formed during the course of these dynamics (Fig. 10.13). However, they can never replace the animated visualisation of the trajectories obtained in MD simulations. Several more- or

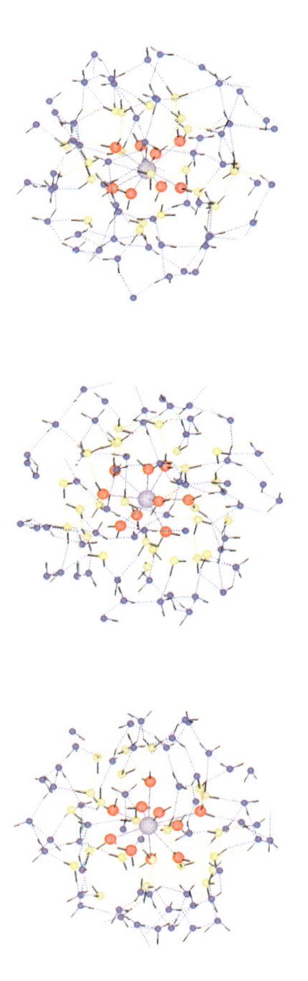

Fig. 10.13 Snapshot pictures taken during MD simulation of Sr^{2+} in water, illustrating some of the different microspecies formed during a time span of 15 ps. First shell coordination numbers are 8, 9 and 10 (from top to bottom).

less-sophisticated software packages have been developed, therefore, to produce a realistic view of chemical processes on the basis of simulations. By using the software package MOLVISION, numerous examples have been provided to illustrate the tool of animated visualisations, and these can be found on the webpage www.molvision.com, from where the reader can download video clips as a supplement to the descriptions given in this chapter.

10.4.3
Specific Evaluations in Macromolecule Simulations

In simulations of large biomolecules, further data treatment is employed to display specific results. One important feature is the root-mean-square-deviation (*rmsd*), corresponding to the sum of squares of all distance deviations along the simulation with respect to a reference structure (mostly the starting or equilibrium structure). The higher the rmsd value, the more changes occur during the simulation. Figure 10.14 shows the rmsd values of an oligonucleotide MD simulation as a function of time. The steadily rising red line – corresponding to a system without consideration of solvent! – indicates that the structure is unstable: the two strands of the nucleic acid uncoil due to the wrong simulation setting. The blue line shows the rmsd value of the same system with a proper simulation setup including the embedding in water: the rmsd value stabilises after a few

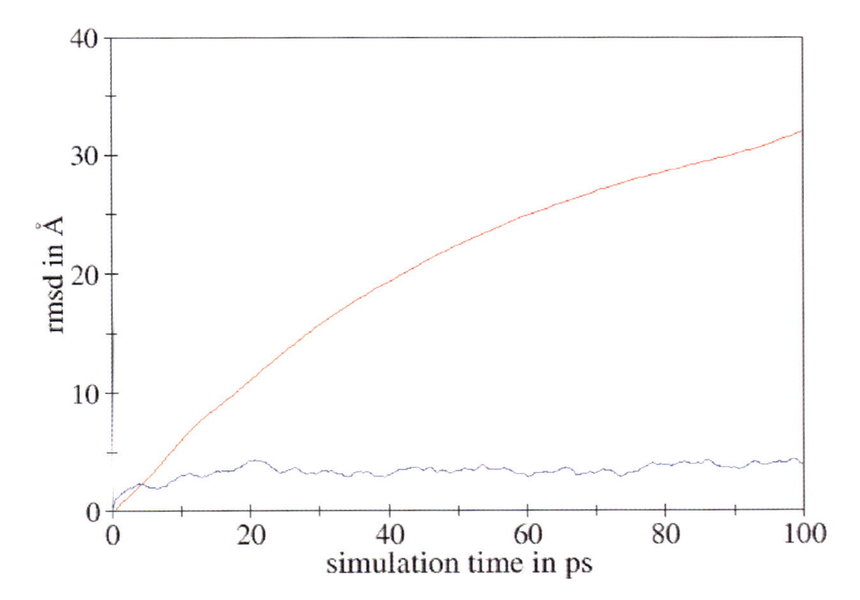

Fig. 10.14 rmsd values of a MD simulation of an oligonucleotide. The red line clearly shows instability, whereas the blue line, fluctuating around a mean value, indicates stabilisation of the DNA structure after a short equilibration.

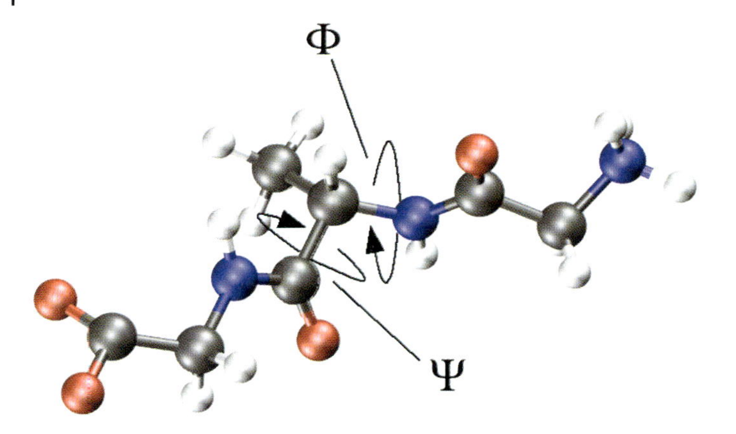

Fig. 10.15 Definition of the angles ϕ and ψ for a Ramachandran plot.

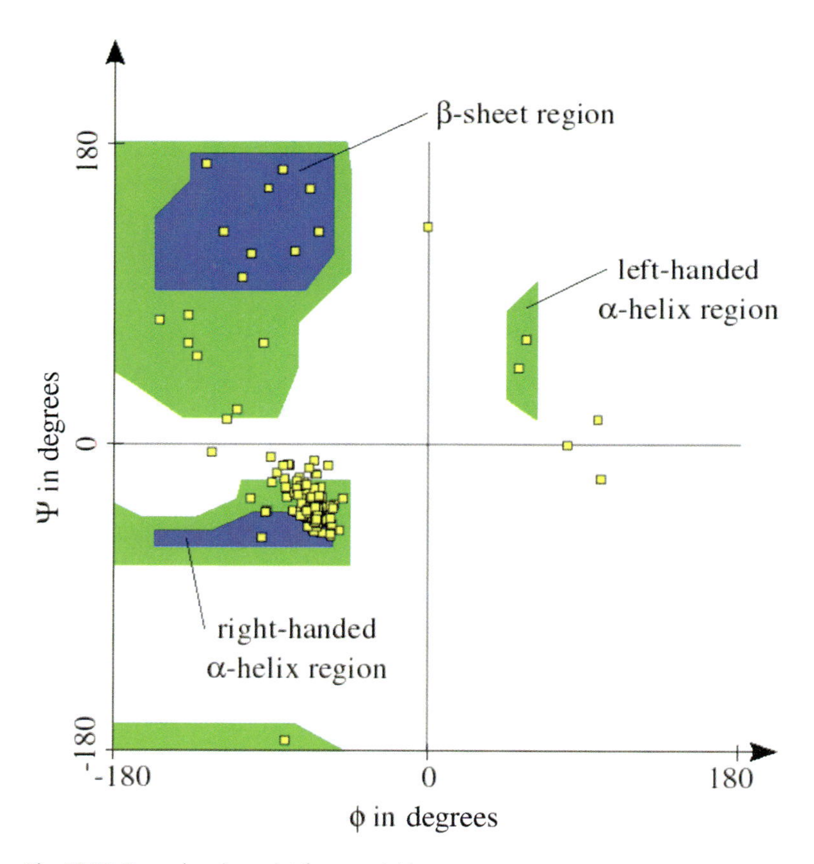

Fig. 10.16 Ramachandran plot for myoglobin.

picoseconds, fluctuating around a mean value. In some cases it proves useful to monitor rmsd values of specific atoms or residues of special interest in order to gain information about the structural and dynamical behaviour of particular regions of the biomolecules. The rmsd values can also be employed for comparisons with experimentally determined structures.

Another helpful macromolecule tool, developed for the analysis of peptides and proteins, is the *Ramachandran plot*. This plot aims at defining stable conformations by varying the ϕ and ψ angles (as defined in Fig. 10.15) in a peptide/protein backbone, assuming planarity for the peptide bonds.

In this method, the atoms of the amino acids are assumed to be hard spheres with the size of the van der Waals radii. Upon variation of the ϕ and ψ angles, distinct areas (the blue regions in Fig. 10.16) can be assigned to the standard structural configurations such as α-helices and β-sheets. If the van der Waals radii are reduced, the amino acids gain more freedom in the configurational space, as depicted by the green regions in Fig. 10.16.

Furthermore, the actual values for ϕ and ψ of each residue of a protein can be displayed in these plots (the yellow markers in Fig. 10.16), providing important information about the structural properties, as the majority of all possible protein conformations can be excluded on the basis of this plot. For example, in the myoglobin molecule (Fig. 10.17) α-helices are dominant, and this can be clearly

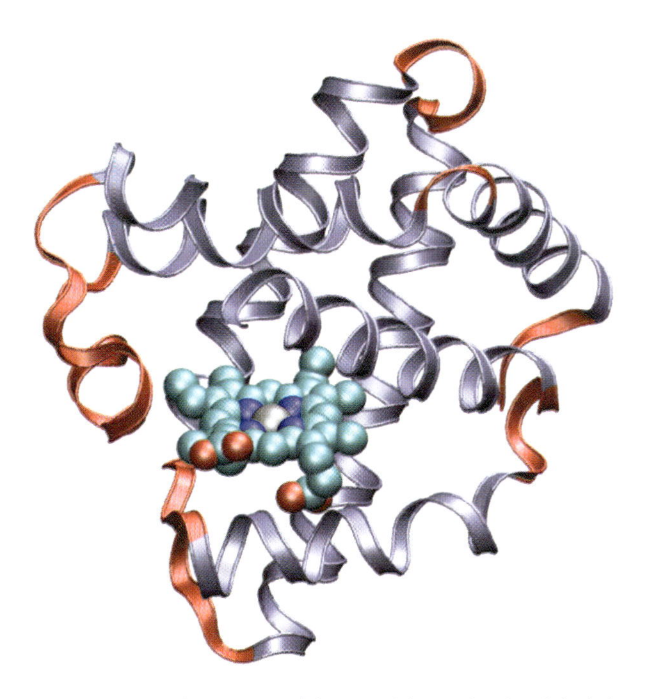

Fig. 10.17 Snapshot picture of the myoglobin molecule: alpha-helices are coloured blue, other regions red.

recognised from the Ramachandran plot of this molecule (Fig. 10.16). Some other residues, mainly located at the terminal parts or in loops, are found in other regions of the plot.

10.5
Quantum Mechanical Simulations

It has been repeatedly stated that the accuracy – and thus the quality – of simulations depends heavily on the potential functions used to calculate the energy (MC) or the forces (MD). Even if the corresponding potential functions have been constructed on the basis of *ab initio*-calculated energy surfaces for pair and 3-body interactions, the accuracy may not be sufficient for a 'flawless' description of a chemical system, in particular a liquid or a solution. The logical solution to this problem would be to evaluate all energies/forces by quantum mechanical methods in every step of the simulation, which would automatically consider all mutual polarisation effects, occurring charge transfers, and higher n-body effects. However, a short estimation of the necessary computational effort for such a procedure shows that this would exceed by far all computer resources presently available. Therefore, a compromise must be sought, and the parameters to adjust the computational effort to an affordable extent are the accuracy of the quantum mechanical formalism and the size of the elementary box. This leads to a typical 'Scylla and Charybdis' situation, where one must navigate between sufficient accuracy (as determined by quantum chemical methodology and box size) and a reasonable computing time. For the latter aspect, the continuously increasing performance of computers in terms of processor speed and lower cost allows quite some optimism for future developments, although even with a 100-fold increased performance of today's best computers, the computational chemist will still encounter serious limits in the application of quantum mechanical simulations. Although parallel processing is helpful, it is not a full remedy, as not all steps of the simulations – in particular of the quantum mechanical part – are 100% parallelisable.

In this situation, another compromise solution has been developed and has proven very successful so far. Instead of reducing the accuracy of the quantum mechanical method and/or the size of the elementary box – both of which would seriously deteriorate the simulation quality – the *hybrid methods* split the elementary box into two (or more) regions. The 'inner' region which contains the chemically most relevant part of the system, and whose properties should be reproduced with the highest possible quality, is treated by quantum mechanics at a sufficiently sophisticated level, whereas the remaining part of the elementary box is treated by conventional, potential-based molecular mechanics. These methods are usually summarised by the acronym 'QM/MM', in connection with the type of simulation performed (i.e., as *QM/MM MC* or *QM/MM MD simulations*).

This approach was first applied to large biomolecules, where the active centre required a better description than force fields could provide, and it is still widely used in this context. In the meantime, a similar methodology has been developed for simulations of liquid systems and solutions, making the QM/MM simulations rather a universal tool, supplying highly accurate data for the liquid state.

A second way to seek a compromise between accuracy and computational effort has been proposed in the form of MD simulations of relatively small molecular clusters based on density functional calculations of the forces. Both methods will be outlined in more detail in the following section.

10.5.1 *Ab initio* QM/MM Simulations

The QM/MM formalism, with its separation of the simulated system into QM and MM regions, can be applied both in MC and MD simulations. In the latter, it is still more demanding in terms of computer time, as a quantum mechanical calculation of forces must be performed in addition to the energy calculation. Despite this disadvantage, most QM/MM studies are conducted in the field of molecular dynamics in order to obtain access to time-dependent data. As mentioned above, the original application of the method was its implementation in simulations of biomolecules. In such cases, less sophistication for the quantum mechanical part is needed, as long as these molecules consist of organic components, for which even semiempirical MO methods may deliver reasonable results. If a metal centre is involved, or if a more refined treatment of weak interactions is needed, then density functional methods can be the method of choice. However, the quality of the results depends on the choice of a suitable functional, and the best functionals usually do not help to save computer time in comparison with *ab initio* HF or even MP/2 methods.

The feasibility of *ab initio* QM/MM simulations is strongly determined by the size of the QM region, the level of theory, and the choice of the basis sets employed. As a rough rule one can state that – depending on the strength and type of interactions within the QM region – a single determinantal approach with a larger QM region is preferable to a higher level of theory restricted to a very small QM region. Furthermore, experience has shown that the choice of basis sets is a most crucial point, and strongly determines the reliability of the simulation results. Double-zeta basis sets seem to be the minimal requirement, augmented in most cases by polarisation and/or diffuse functions.

A brief outline of the QM/MM MD formalism will give us the opportunity to discuss some further details of the methodical problems. Figure 10.18 illustrates the separation scheme of a typical QM/MM simulation, showing the spherical QM region within the cubic elementary box.

In this scheme the forces are calculated according to

$$F_{tot} = F_{MM}^{sys} + (F_{QM}^{QM} - F_{QM}^{MM}) \cdot S(r),$$

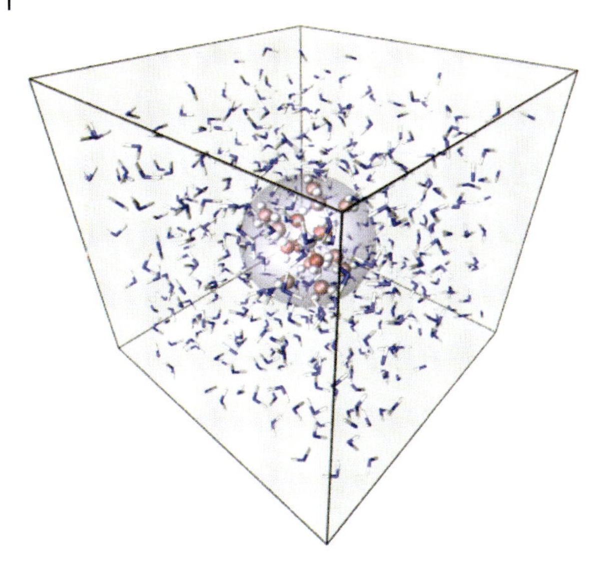

Fig. 10.18 Separation scheme of a typical QM/MM simulation, showing the spherical QM region within a cubic elementary box.

where F_{tot} is the total force of the system, F_{MM}^{sys} is the MM force of the whole system, F_{QM}^{QM} the QM force in the QM region, and F_{QM}^{MM} the MM force of the QM region. F_{QM}^{MM} accounts for the coupling between the QM and MM regions. To ensure a smooth transition and continuous change of forces between the QM and MM regions, a smoothing function $S(r)$ should be applied between the radii r_{on} and r_{off}, the difference of which is usually 0.2 Å. This smoothing function is defined as

$$S(r) = \begin{cases} 0 & \text{for } r < r_{on} \\ \frac{(r_{off}^2 - r^2)^2 (r_{off}^2 + 2r^2 - 3r_{on}^2)}{(r_{off}^2 - r_{on}^2)^3} & \text{for } r_{on} < r < r_{off} \\ 1 & \text{for } r > r_{off} \end{cases}$$

and is of particular importance when particles move between the QM and MM regions during the course of the simulation (larger molecules transiting the border imply more difficulties in achieving smooth transitions).

Another problem which is clearly recognisable from the picture in Fig. 10.18 arises in the treatment of large biopolymers, where the QM region is only a subregion of this molecule. In this case, the separation of the QM and MM regions will cut through covalent bonds, unlike a liquid, where relatively small and weakly interacting molecules are either inside or outside the QM region. To overcome this problem, the covalent bonds affected by the separation can be broken and the open valences closed by 'dummy' atoms (usually hydrogens), at the price of a significant change of the chemical system investigated.

All types of QM/MM simulation are still extremely time-demanding, lasting from a few weeks to several months on a multi-processor, high-performance computer. The simulation time covered by this effort will still be in the picosecond range (10–50 ps), if the time step width is selected in the sub-femtosecond range to allow for explicit movements of hydrogen atoms. On the other hand, the correct evaluation of dynamical data apparently requires the accuracy of *ab initio* simulations, as many examples have recently shown. As already indicated, however, the rapid development of computer technology and speed will most certainly lead to a rapid improvement of the situation. Two decades ago, classical MD simulations were still a major task for the best 'supercomputers', but today they can be performed overnight on a good personal computer. One can expect, therefore, that in the near future *ab initio* simulations will become possible for systems with a large QM region and at higher levels of theory, using standard computational equipment.

Some recent methodical improvements of QM/MM simulations have also paved the way for the simulation of more complex compounds, and without the need of any potential functions except that for solvent–solvent interactions. The *Quantum Mechanical Charge Field (QMCF) MD* method is an example, where this improvement is achieved by a substantial extension of the QM region, a continuous consideration of the fluctuating charges within this region, and an incorporation of the charges in the MM region as a perturbational part of the Hamiltonian.

10.5.2 Car–Parrinello DFT Simulations

The quest for a higher accuracy than classical potentials can achieve, and the introduction of quantum mechanics into the simulation procedure, has produced another methodical approach, known as *Car–Parrinello simulations*. In this method, the total system is treated by DFT, and the compromise to maintain the computing time manageable is twofold. On the one hand, the number of molecules contained in the elementary box is reduced to 30–60, while on the other hand the DFT treatment is usually limited to a simple GGA-type functional (mostly BLYP or PBE), thus giving the quantum mechanical treatment a semi-empirical nature (though the DFT community still likes to call this framework 'first principle' or '*ab initio*' simulations). Besides the quantum mechanical accuracy problem and the difficulty of a method-inherent quality control, the size of the system is the main problem of this type of simulation. If one considers a larger ion in solution, the number of solvent molecules provided can be just sufficient or even too small to form the complete solvation shells, and nothing is left to describe the embedding of the solvate in bulk solvent, or any exchange processes between solvation shells and bulk. This problem may be overcome by the expected increase in computational power, allowing the use of larger numbers of molecules, which would also prevent unrealistic symmetry effects when applying periodic boundary conditions. Even then, the implementation of more suitable density functionals will further increase the computational demand. A real breakthrough in DFT could provide a new dimension for this method, though it does not yet seem to have appeared on the horizon.

Test Questions Related to this Chapter

1. How is temperature controlled in MC and MD simulations?
2. Which factors determine the size of an elementary box?
3. How can we ensure that the simulation reflects the properties of a bulk liquid?
4. How can we control, whether a simulation has reached the best possible equlibrium state of the system?
5. What properties of the simulated system can be extracted from RDFs?
6. How large should one choose the time step in a MD simulation of a hydrogen-bonded liquid?
7. Why do we store velocities in a trajectory file?
8. How can we construct analytical functions for intermolecular interactions, and what terms are most common in such functions?
9. How can we implement quantum mechanics in a statistical simulation?
10. What kind of potentials are used for the simulation of biopolymers?

11 Synopsis

γηρασκω αει διδασκομενος – I am aging, continuously being taught.

Sokrates 5th century BC – 2400 years before the term 'life long learning' was 'invented'.

The ten chapters of this book are aimed at introducing modern theoretical and computational chemistry knowledge 'in a nutshell'. The character of the book, in attempting to concentrate on the most basic and relevant features, implies that none of the topics could be treated in an exhaustive and complete way. However, the reader who has become familiar with the book's contents has certainly acquired a sound background in contemporary theoretical chemistry, and will be able critically to evaluate the current models being used in chemistry and publications involving computational chemistry methods. The book should also serve as a sufficient preparation for more specialised reading in the various fields of theoretical and computational chemistry – and hopefully serve as a 'starter', evoking interest for further involvement in these fields.

It is also hoped that experimental chemists will make use of this book to update themselves on the validity of common chemical models, and the manifold possibilities theoretical chemistry provides nowadays to improve and supplement the interpretation of experimental data. This 'bridging of the gap' between experimental and theoretical chemistry – which seems mandatory for modern chemical education and research – has been one of the most desired targets of this book. The compactness of the book and its 'bandwith' in terms of the subjects presented should encourage also those unlikely to make theoretical chemistry their main subject. In this way the subject will achieve the popularity it deserves among the entire chemical community.

Theoretical chemistry has – through its computational methods – gained much importance and relevance, not only among the scientific community, but also in practical applications. This advance is to be seen only as a beginning, however, indicating a fundamental reformation process of chemistry and other sciences built on chemical knowledge and processes. Consequently, the niches where one can work with conventional chemical methods will become narrower and increasingly irrelevant.

The Basics of Theoretical and Computational Chemistry. Edited by B. M. Rode, T. S. Hofer and M. D. Kugler

ISBN: 978-3-527-31773-8

Let us return to the introductory sentence of this book, *"Every experimental result is – at the best – as good as the theoretical model used for its interpretation ..."*. During the course of this book it should have become clear that many chemists are still using models that are incompatible with modern theory, and that a revision of many well-established interpretations appears unavoidable. It is hoped that this book, in attempting to present theoretical chemistry in a short and mathematically simple way, could provide some basis and inspiration for the process of modernising chemistry. At the same time, it should encourage experimental and theoretical chemists to join forces, much as occurred in physics a century ago, leading to unprecedented achievements. The central importance of Schrödinger's equation as the quintessence of chemical theory has been explicitly and implicitly demonstrated in relation to foundations and applications of modern chemical theory. This importance has even become the theme of a modern song, composed by the Austrian musician, 'Starmania' winner and pop-star Michael Tschuggnall (who is a student of computer science), called '*Schrödingers equation*', which can be heard on the webpage http://www.theochem.at.

Appendix 1
Ab Initio Hartree–Fock Calculations for Hyposulfuric Acid (H_2SO_3), including Optimisation of the Geometry

Ab initio HF calculation with STO 6-31G(d) basis set
Program: GAUSSIAN 03
INPUT DATA
Charge = 0 Multiplicity = 1

Atoms and Cartesian Coordinates

S	−0.20974	0.03262	0.3886
O	0.34406	1.41884	−0.31796
O	−1.23348	−0.60876	−0.46062
O	1.20901	−0.8436	0.09158
H	0.9396	−1.74023	−0.19036
H	0.64197	1.23748	−1.23429

Distance matrix (Å):

		1	2	3	4	5
1	S	0.000000				
2	O	1.651523	0.000000			
3	O	1.476681	2.572964	0.000000		
4	O	1.693763	2.456521	2.515121	0.000000	
5	H	2.190702	3.217246	2.464861	0.977761	0.000000
6	H	2.193369	0.980460	2.743079	2.531869	3.169405

SCF-Calculation

68 basis functions, 144 primitive gaussians, 68 cartesian basis functions
21 alpha electrons 21 beta electrons
nuclear repulsion energy 190.0769018146 Hartrees.
SCF Done: E(RHF) = −623.178524267 A.U.

The Basics of Theoretical and Computational Chemistry. Edited by B. M. Rode, T. S. Hofer and M. D. Kugler

ISBN: 978-3-527-31773-8

Gradient calculation

Center number	Atomic number	Forces (Hartrees/Bohr)		
		X	Y	Z
1	16	0.008342997	−0.011756283	−0.044191108
2	8	−0.005127655	−0.034623880	−0.010114390
3	8	0.037123811	0.022931330	0.026283490
4	8	−0.038448970	−0.003061191	−0.000111011
5	1	0.006122828	0.022432060	0.005649024
6	1	−0.008013012	0.004077964	0.022483995

Convergence of gradient calculation

Item	Value	Threshold	Converged?
Maximum Force	0.000276	0.000450	YES
RMS Force	0.000094	0.000300	YES
Maximum Displacement	0.001496	0.001800	YES
RMS Displacement	0.000733	0.001200	YES

Predicted change in Energy = −1.670369D-07
Optimization completed.
– Stationary point found.

SCF cycle for optimised geometry

E = −623.188272263192
E = −623.188274462257 Delta-E = −0.000002199065
E = −623.188274701898 Delta-E = −0.000000239641
E = −623.188274725587 Delta-E = −0.000000023689
E = −623.188274729356 Delta-E = −0.000000003769
E = −623.188274729777 Delta-E = −0.000000000421
E = −623.188274729810 Delta-E = −0.000000000033
E = −623.188274729815 Delta-E = −0.000000000005
E = −623.188274729816 Delta-E = −0.000000000001

Final results

Coordinates (cf. Fig. 8.1)

Center number	Atomic number	Atomic type	Coordinates (Å)		
			X	Y	Z
1	16	0	−0.161931	0.044270	0.418623
2	8	0	0.507954	1.281516	−0.354132
3	8	0	−1.127329	−0.501311	−0.508201
4	8	0	1.062827	−0.993043	0.406496
5	1	0	1.033920	−1.523325	−0.388987
6	1	0	0.375978	1.188243	−1.296848

Distance matrix (Å):

		1	2	3	4	5
1	S	0.000000				
2	O	**1.605203**	0.000000			
3	O	**1.445217**	2.424120	0.000000		
4	O	**1.605054**	2.461718	2.423892	0.000000	
5	H	2.130645	2.853942	2.393684	**0.956466**	0.000000
6	H	2.130930	**0.956468**	2.395097	2.851517	2.934229

Energy

E(RHF) = −623.188274730 A.U.

Eigenvalues

Alpha occ. eigenvalues –	−92.18049	−20.62976	−20.62971	−20.56062	−9.16161
Alpha occ. eigenvalues –	−6.84437	−6.84428	−6.84078	−1.50773	−1.40709
Alpha occ. eigenvalues –	−1.36408	−0.94360	−0.77824	−0.72121	−0.69196
Alpha occ. eigenvalues –	−0.66261	−0.58834	−0.55272	−0.54043	−0.48693
Alpha occ. eigenvalues –	−0.43673				

....

Alpha virt. eigenvalues –	0.17687	0.17990	0.22841	0.25501	0.33779
Alpha virt. eigenvalues –	0.44537	0.51573	0.52119	0.61298	0.84611
Alpha virt. eigenvalues –	0.86513	0.92994	1.03116	1.03386	1.07797

Population analysis

Condensed to atoms (all electrons):

		1	2	3	4	5	6
1	S	13.989910	0.116814	0.414292	0.116869	0.012545	0.012594
2	O	0.116814	8.507404	−0.051518	−0.034258	−0.001720	0.236735
3	O	0.414292	−0.051518	8.410826	−0.051533	0.007650	0.007619
4	O	0.116869	−0.034258	−0.051533	8.507392	0.236797	−0.001724
5	H	0.012545	−0.001720	0.007650	0.236797	0.269881	0.001121
6	H	0.012594	0.236735	0.007619	−0.001724	0.001121	0.270023

Mulliken atomic charges:

1	S	1.336976
2	O	−0.773456
3	O	−0.737336
4	O	−0.773543
5	H	0.473727
6	H	0.473632

Sum of Mulliken charges = 0.00000

Dipole moment (field-independent basis, Debye):

X = 0.0050   Y = −0.8818   Z = 1.4969   Tot = 1.7373

Quadrupole moment (field-independent basis, Debye-Ang):

XX = −28.0383   YY = −36.1819   ZZ = −24.7231
XY = −0.0031   XZ = −0.0026   YZ = −2.4473

Appendix 2
Books Recommended for Further Reading

Andrew R. Leach, *Molecular Modelling – Principles and Applications.* Pearson Educaton Ltd., Second Edition 2001, (ISBN: 0-582-38210-6).

Frank Jensen, *Introduction to Computational Chemistry.* John Wiley and Sons Ltd., Second Edition 2006, (ISBN: 0-470-01187-4).

Christoph J. Cramer, *Essentials of Computational Chemistry – Theories and Models.* John Wiley and Sons Ltd., Second Edition 2004, (ISBN: 0-470-09181-9).

Albert Messiah, *Quantum Mechanics.* Dover Publications, 2000, (ISBN: 0-486-40924-4).

Attila Szabo and Neil S. Ostlund, *Modern Quantum Chemistry – Introduction to Advanced Electronic Structure Theory.* Dover Publications, 1996, (ISBN: 0-486-69186-1).

Trygve Helgaker, Paul Jørgensen, and Jeppe Olsen, *Molecular Electronic-Structure Theory.* John Wiley and Sons Ltd., 2000, (ISBN: 0-471-96755-6).

Stephen Wilson, *Electron Correlation in Molecules.* Oxford University Press, 1985, (ISBN: 0-198-55617-9).

Wolfram Koch and Max C. Holthausen, *A Chemist's Guide to Density Functional Theory.* John Wiley and Sons Ltd., 2000, (ISBN: 3-527-30372-3).

Mike P. Allen and Dominic J. Tildesley, *Computer Simulation of Liquids.* Clarendon Press, 1989, (ISBN: 0-198-55645-4).

Daan Frenkel and Berend Smit, *Understanding Molecular Simulations – From Algorithms to Applications.* Academic Press, Second Edition 2001, (ISBN: 0-122-67351-4).

David M. Bishop, *Group Theory and Chemistry.* Dover Publications, 1993, (ISBN: 0-486-67355-3).

The Basics of Theoretical and Computational Chemistry. Edited by B. M. Rode, T. S. Hofer and M. D. Kugler

ISBN: 978-3-527-31773-8

Index

The Basics of Theoretical and Computational Chemistry. Edited by B. M. Rode, T. S. Hofer and M. D. Kugler

ISBN: 978-3-527-31773-8

d

e

f

g

Related Titles

Matta, C. F., Boyd, R. J. (eds.)

The Quantum Theory of Atoms in Molecules

From Solid State to DNA and Drug Design

2007

ISBN-13: 978-3-527-30748-7
ISBN-10: 3-527-30748-6

Dronskowski, R.

Computational Chemistry of Solid State Materials

A Guide for Materials Scientists, Chemists, Physicists and others

2005

ISBN-13: 978-3-527-31410-2
ISBN-10: 3-527-31410-5

Cramer, C. J.

Essentials of Computational Chemistry

Theories and Models

2004

ISBN-13: 978-0-470-09182-1
ISBN-10: 0-470-09182-7

Gasteiger, J., Engel, T. (eds.)

Chemoinformatics

A Textbook

2003

ISBN-13: 978-3-527-30681-7
ISBN-10: 3-527-30681-1

Höltje, H.-D., Sippl, W., Rognan, D., Folkers, G.

Second Edition

Molecular Modeling

Basic Principles and Applications

2003

ISBN-13: 978-3-527-30589-6
ISBN-10: 3-527-30589-0

Kutzelnigg, W.

Einführung in die Theoretische Chemie

2003

ISBN-13: 978-3-527-30609-1
ISBN-10: 3-527-30609-9

Harrison, P.

Computational Methods in Physics, Chemistry and Biology

An Introduction

2001

ISBN-13: 978-0-471-49563-5
ISBN-10: 0-471-49563-8

Koch, W., Holthausen, M. C.

A Chemist's Guide to Density Functional Theory

2001

ISBN-13: 978-3-527-30422-6
ISBN-10: 3-527-30422-3